BOREAL BIRDS *of* NORTH AMERICA

STUDIES IN AVIAN BIOLOGY

A Publication of the Cooper Ornithological Society

WWW.UCPRESS.EDU/GO/SAB

Studies in Avian Biology is a series of works published by the Cooper Ornithological Society since 1978. Volumes in the series address current topics in ornithology and can be organized as monographs or multi-authored collections of chapters. Authors are invited to contact the Series Editor to discuss project proposals and guidelines for preparation of manuscripts.

See complete series list on page 135.

BOREAL BIRDS *of* NORTH AMERICA

A Hemispheric View of Their Conservation Links and Significance

Jeffrey V. Wells, *Editor*

Studies in Avian Biology No. 41

A PUBLICATION OF THE COOPER ORNITHOLOGICAL SOCIETY

University of California Press

Berkeley   Los Angeles   London

University of California Press, one of the most distinguished university presses in the United States, enriches lives around the world by advancing scholarship in the humanities, social sciences, and natural sciences. Its activities are supported by the UC Press Foundation and by philanthropic contributions from individuals and institutions. For more information, visit www.ucpress.edu.

Studies in Avian Biology, No. 41
For digital version, see the UC Press website.

University of California Press
Berkeley and Los Angeles, California

University of California Press, Ltd.
London, England

Library of Congress Cataloging-in-Publication Data

Boreal birds of North America : a hemispheric view of their conservation links and significance / Jeffrey V. Wells, editor.
   p. cm.—(Studies in avian biology ; no. 41)
"A publication of the Cooper Ornithological Society."
Includes bibliographical references and index.
ISBN 978-0-520-27100-5 (cloth : alk. paper)
1. Forest birds—Canada. 2. Forest birds—Alaska. 3. Birds—Ecology—North America.
4. Taiga ecology—North America. I. Wells, Jeffrey V. (Jeffrey Vance), 1964–.

QL685.B67   2011
639.9'78—dc23
                                                                  2011028666

19 18 17 16 15 14 13 12 11
10 9 8 7 6 5 4 3 2 1

The paper used in this publication meets the minimum requirements of ANSI/NISO Z39.48-1992 (R 1997) (Permanence of Paper).

Cover image: Pine Grosbeak (*Pinicola enucleator*). Photo © Jeff Nadler.

PERMISSION TO COPY

DEDICATION

To the indigenous peoples of the Boreal Forest region,
whose stewardship has maintained harmony with the birds
and the land for thousands of years.

And to Carl Marti, who began the process to bring this book
forward though the Studies in Avian Biology series but was lost
to the ornithological community before the volume was
completed.

CONTENTS

CONTRIBUTORS

MICHAEL L. AVERY
Wildlife Services
National Wildlife Research Center
2820 E. University Avenue
Gainesville, FL 32641, USA
michael.l.avery@aphis.usda.gov

LOUIS BEVIER
Department of Biology
5720 Mayflower Hill Drive
Colby College
Waterville, ME 04901, USA
lrbevier@colby.edu

PETER J. BLANCHER
Environment Canada
National Wildlife Research Centre
Ottawa, ON K1A 0H3, Canada
peter.blancher@ec.gc.ca

W. SEAN BOYD
Canadian Wildlife Service
5421 Robertson Road
Delta, BC V4K3N2, Canada
sean.boyd@ec.gc.ca

ERIC W. BUTTERWORTH
Western Boreal Office
Ducks Unlimited Canada
100, 17958 – 106 Avenue
Edmonton, AB T5S 1V4, Canada
e_butterworth@ducks.ca

ROB CLAY
BirdLife International
17-17-717 Casilla Postal
Quito, Ecuador
rob.clay@birdlife.org.ec

ANDREW R. COUTURIER
Bird Studies Canada
P.O. Box 160, 115 Front Street
Port Rowan, ON N0E 1M0, Canada
acouturier@bsc-eoc.org

IAN J. DAVIDSON
Nature Canada
75 Albert Street, Suite 300
Ottawa, ON K1P 5E7, Canada
idavidson@naturecanada.ca

SUSAN W. DE LA CRUZ
USGS Western Ecological Research Center
San Francisco Bay Estuary Field Station
505 Azuar Drive
Vallejo, CA 94592, USA, and
Graduate Group in Ecology
University of California
1 Shields Avenue
Davis, CA 95616, USA
susan_delacruz@usgs.gov

DEAN W. DEMAREST
U.S. Fish and Wildlife Service
1875 Century Boulevard, Suite 240
Atlanta, GA 30345, USA
dean_demarest@fws.gov

JOHN M. EADIE
Department of Fisheries, Wildlife,
and Conservation Biology,
Avian Sciences
Graduate Group
1 Shields Avenue
Davis, CA 95616, USA
jmeadie@ucdavis.edu

DANIEL ESLER
Canadian Wildlife Service
5421 Robertson Road
Delta, BC V4K3N2, Canada
desler@sfu.ca

JOSEPH R. EVENSON
Washington Department
of Fish and Wildlife
7801 Phillips Road
Lakewood, WA 98498, USA
evensjre@dfw.wa.gov

DAVID EVERS
BioDiversity Research Institute
19 Flaggy Meadow Road
Gorham, ME 04038, USA
david.evers@briloon.org

DAVID D. FERNÁNDEZ
BirdLife International
17-17-717 Casilla Postal
Quito, Ecuador
david.diaz@birdlife.org.ec

RUSSELL GREENBERG
Smithsonian Migratory Bird Center
National Zoological Park
Washington, DC 20008, USA
greenbergr@si.edu

PAUL B. HAMEL
USDA Forest Service
Southern Hardwoods Laboratory
P.O. Box 227
Stoneville, MS 38776, USA
phamel@fs.fed.us

KEITH A. HOBSON
Canada Canadian Wildlife Service
115 Perimeter Road
Saskatoon, SK S7N 0X4, Canada
keith.hobson@ec.gc.ca

A DRIENNE J. LEPPOLD
University of Maine, Ecology and
Environmental Sciences
217 Murray Hall
Orono, ME 04473, USA
aleppold@gmail.com

JASON LUSCIER
Department of Biological Sciences
University of Arkansas
Fayetteville, AR 72701, USA
jluscie@uark.edu

RICH MACDONALD
The Natural History Center
6 Firefly Lane
P.O. Box 6
Bar Harbor, ME 04609, USA
rich@thenaturalhistorycenter.com

GLENN G. MACK
Western Boreal Office
Ducks Unlimited Canada
100, 17958 – 106 Avenue
Edmonton, AB T5S 1V4, Canada
g_mack@ducks.ca

STEVEN M. MATSUOKA
U.S. Fish and Wildlife Service
1011 East Tudor Road
Anchorage, AK 99503, USA
steve_matsuoka@fws.gov

CLAUDIA METTKE-HOFMANN
School of Natural Sciences and Psychology
Liverpool John Moores University
James Parsons Building
Byrom Street
Liverpool L3 3AF, United Kingdom
mettke@erl.orn.mpg.de

JULIENNE L. MORISSETTE
Western Boreal Office
Ducks Unlimited Canada
100, 17958 – 106 Avenue
Edmonton, AB T5S 1V4, Canada
j_morissette@ducks.ca

ROBERT S. MULVIHILL
Carnegie Museum of Natural History
Powdermill Avian Research Center
1847 Route 381
Rector, PA 15677, USA
robert.mulvihill@gmail.com

DANIEL K. NIVEN
National Audubon Society
Illinois Natural History Survey
607 East Peabody Drive
Champaign, IL 61820, USA
dniven@audubon.org

DAVID R. NYSEWANDER
Washington Department of Fish and Wildlife
600 Capitol Way North
Olympia, WA 98501, USA
drn2inoly@comcast.net

ERIC C. PALM
U. S. Geological Survey
Western Ecological Research Center
San Francisco Bay Estuary Field Station
505 Azuar Drive
Vallejo, CA 94592, USA
epalm@sfu.ca
(Current address:
Center for Wildlife Ecology
Department of Biological Sciences
Simon Fraser University
Burnaby, BC V5A 1S6, Canada)

LUKE L. POWELL
University of Maine
Department of Biological Sciences
5751 Murray Hall, Rm. 207
Orono, ME 04469, USA
luke.l.powell@gmail.com

BRUCE A. ROBERTSON
Smithsonian Conservation
Biology, Institute
Migratory Bird Center
National Zoological Park
Washington, DC 20013, USA
brucerobertson@hotmail.com

DAVID SHAW
Alaska Bird Observatory
418 Wedgewood Drive
Fairbanks, AK 99701, USA
dshaw@alaskabird.org

STUART M. SLATTERY
Institute for Wetland and Waterfowl Research
Ducks Unlimited Canada
Box 1160
Stonewall, MB R0C 2Z0, Canada
s_slattery@ducks.ca

JOHN Y. TAKEKAWA
U. S. Geological Survey
Western Ecological Research Center
San Francisco Bay Estuary Field Station
505 Azuar Drive
Vallejo, CA 94592, USA
john_takekawa@usgs.gov

DAVID H. WARD
U. S. Geological Survey
Alaska Science Center
4210 University Drive
Anchorage, AK, USA
david_ward@usgs.gov

JEFFREY V. WELLS
Boreal Songbird Initiative
1904 Third Avenue, Suite 305
Seattle, WA 98101 USA
jwells@intlboreal.org

MATTHEW T. WILSON
Department of Fisheries, Wildlife, and
Conservation Biology
Avian Sciences Graduate Group
1 Shields Avenue
Davis, CA 95616, USA
matt_wilson@fws.gov
(Current address:
Stillwater National Wildlife Refuges
100 Auction Road
Fallon, NV 89406, USA)

JULIE YEE
Western Ecological Research Center
3020 State University Drive East
Modoc Hall
Sacramento, CA 95819, USA
julie_yee@usgs.gov

PREFACE

Stretching across 1.5 billion acres of Canada and Alaska, North America's boreal is the world's largest remaining intact primary forest. This vast ecosystem is one of the last global havens for large herds of migratory caribou and teems with moose, wolves, and grizzly bears. Uncountable pristine lakes, rivers, and wetlands harbor equally robust freshwater fisheries. And, not surprisingly, this vast refuge for nature provides critical habitat for billions of birds.

The majority of birds found in the boreal forests of Canada and Alaska migrate south after the breeding season, where they appear as seasonal visitors to millions of people across the United States, Mexico, the Caribbean, and Central and South America. In order to preserve the spectacle and miracle of these annual sojourns, all nations along the migratory route must act as stewards of the passageway connecting their habitats.

Canada has taken important, world-leading steps in the effort to ensure that migrating birds will have ample and appropriate nesting and rearing habitat. Aboriginal governments and communities throughout Canada have developed land-use plans specifically to protect habitat for birds and other wildlife. The provinces of Ontario and Quebec have pledged to work with First Nations to conserve hundreds of millions of acres of boreal habitat, and in the Northwest Territories, the federal government has worked with aboriginal communities to establish tens of millions of acres of protected areas.

Thanks to these visionary efforts, the future of the North American boreal forest looks promising, but Canada and the United States have more work to do, and to that end, we must continue to learn more about this great forest and its inhabitants. This book does much to advance our understanding of boreal bird biology and conservation. I hope that it will serve as a model in exploring new and interesting facets of the lives of the migratory birds that grace our hemisphere.

STEVE KALLICK
Director, International Boreal Conservation Campaign, Pew Environment Group

ACKNOWLEDGMENTS

Steve Kallick of the Pew Environment Group's International Boreal Conservation Campaign; Marilyn Heiman, Lane Nothman, and Gary Stewart of the Boreal Songbird Initiative; Larry Innes of the Canadian Boreal Initiative; and Fritz Reid of Ducks Unlimited have together shown remarkable foresight and leadership in furthering the conservation of North America's boreal forest region and in recognizing the importance of birds in connecting the boreal forest region to the rest of the Americas. I thank them for their support and encouragement in developing the science that portrays these connections and in particular for their support of the symposium at the North American Ornithological Conference, from which this volume was initiated. My sincerest appreciation goes to Mathew Medler, who was instrumental in organizing the symposium and in shepherding the submission of manuscripts and reviews through the first part of this book project. I was fortunate to receive assistance in assembling various items necessary to the publication of this book from Alecia Wells, Jill Jamison, and Jennifer Cerulli. Lauren Mier was kind enough to provide design help under short notice.

I am indebted to the expert reviewers who volunteered time from their busy schedules to review manuscripts for this volume. They included Louis Bevier, Peter Blancher, David Bonter, Stephen Brown, Mike Burger, Dan Lebbin, Daniel Mazerolle, Kevin McGowan, Robert Mulvihill, David Pashley, Chris Rimmer, Dina Roberts, Bruce Robertson, Iain Stenhouse, Stuart Slattery, and John Takekawa.

I especially wish to thank the original Studies in Avian Biology series editor, Carl Marti, for paving the way for the project, and to the current editor, Brett Sandercock, who so ably stepped in when Carl passed away and who has patiently and expertly carried this project to completion. Finally, I wish to thank all the authors for their patience and diligence in completing their chapters and making this book a useful and interesting contribution to our understanding of boreal birds.

JEFFREY V. WELLS
Boreal Songbird Initiative

CHAPTER ONE

Boreal Forest Threats and Conservation Status

Jeffrey V. Wells

North America's boreal forest region is considered one of the most intact and least disturbed of the globe's terrestrial forested ecosystems (Lee et al. 2006). Its nearly 600 million hectares span from interior Alaska across Canada to Labrador and Newfoundland (Fig. 1.1) and encompass some of the world's largest peatlands, lakes, and rivers (Schindler and Lee 2010, Wells et al. 2011), major stores of terrestrial carbon (Carlson et al. 2009, 2010; Tarnocai et al. 2009), large populations of carnivores (Canadian Boreal Initiative 2005, Cardillo et al. 2006, Bradshaw et al. 2009), and some of the world's last remaining unchecked large mammal migrations (Wilcove 2008, Hummel and Ray 2009).

This volume highlights new research that is illuminating the importance of the region to North America's avifauna and the complexity of avian ecological connectivity between the boreal forest region and ecoregions throughout the Americas. The contributions showcase a unique set of perspectives on the migration, wintering ecology, and conservation of avifaunal communities that are tied to the boreal forest in ways that may not have been previously considered.

In North America's boreal forest, as in its Southern Hemisphere counterpart, the Amazon, development and land-use management decisions are occurring at an accelerated rate. An assessment in 1987 suggested that 26% of the "frontier forests" of North America (virtually all in the boreal forest region) were under moderate or high threat (Bryant et al. 1997). Another analysis ranked two southern boreal forest ecoregions as being in Critically Endangered condition, one as Endangered, and seven additional boreal forest ecoregions as Vulnerable (Ricketts et al. 1999)

Estimates of the amount of habitat in the southern boreal forest that is no longer intact range as high as 66% (Ricketts et al. 1999), encompassing 177 million hectares. Using satellite imagery, Lee et al. (2006) documented that less than 15% of the 71 million hectare Boreal Plains ecozone (the portion of the southern boreal extending from the eastern foothills of the Canadian Rockies to south-central Manitoba) remains in large, intact forest landscapes. Between 1990 and 2000 over 400,000 hectares of the southern boreal of Saskatchewan and Manitoba and over 2.4 million hectares of the boreal of Quebec was disturbed by human-caused influences including forestry, road-building, and other infrastructure development (Stanojevic et al. 2006a, 2006b).

Since 1975 over 31 million ha of Canadian forest have been harvested (Canadian Council of Forest Ministers 2010). Between 1990 and 2008 the total area harvested in Canada was 18,412,244 ha (Canadian Council of Forest Ministers 2010).

Wells, J. V. 2011. Boreal forest threats and conservation status. Pp. 1–6 *in* J. V. Wells (editor). Boreal birds of North America: a hemispheric view of their conservation links and significance. Studies in Avian Biology (no. 41), University of California Press, Berkeley, CA.

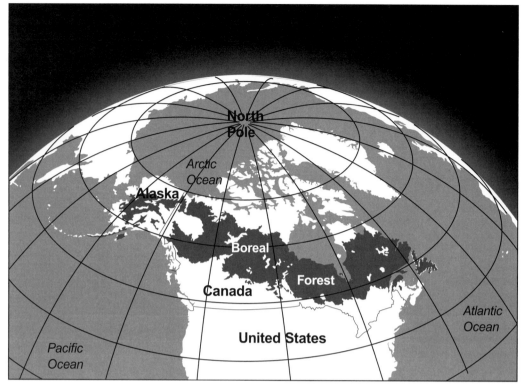

Figure 1.1. Extent of boreal forest in North America.

Assuming the same rate of harvest and that 65% of the Canada's timber harvest occurs in the boreal forest region, about 6 million ha will be harvested in Canada's boreal region over the next ten years.

Many other kinds of industrial disturbances are taking place within the boreal forest region. Oil and gas exploration and extraction activities, especially in the western boreal forest region, are rapidly increasing. In Canada, a record 22,800 oil and gas wells were drilled in 2004, and the number of new wells drilled annually is projected to continue increasing (Canadian Association of Petroleum Producers 2005). The industrial footprint from oil and gas extraction activities throughout Canada's boreal forest region as of 2003 was estimated at 46 million ha, or approximately 8% of Canada's boreal forest region (Anielski and Wilson 2005). Within Alberta's oil sands region, habitat that would have supported an estimated 58,000–402,000 breeding birds has already been lost (Timoney and Lee 2009) and future losses have been projected into the tens of millions (Wells et al. 2008).

In Canada, large hydropower projects developed in the 1970s and 1980s have flooded millions of hectares (Wells et al. 2011), especially in parts of the eastern boreal forest region. For example, five reservoirs established in the La Grande River region of central Quebec flooded 1.1 million hectares of terrestrial habitat (Gauthier and Aubry 1996).

Currently, approximately 70% of bird species that regularly breed in Canada's boreal forest region show impacts from anthropogenic disturbance (road building, forestry, mining, etc.) within at least 10% of their distribution, while in only 11% of bird species is at least 10% of the range within protected areas (J. V. Wells, unpubl. data). For example, 24% of the breeding distribution of Canada Warbler (*Wilsonia canadensis*) in the Canadian boreal forest region is within areas impacted by anthropogenic disturbance, while 7% of its distribution is within protected areas (Fig. 1.2). In the case of the Evening Grosbeak (*Coccothraustes vespertinus*), 39% of its breeding distribution in Canada's boreal forest is within areas impacted by anthropogenic disturbance, while 9% is within protected areas (Fig. 1.3).

Fortunately, great strides have been and continue to be made in the conservation of North

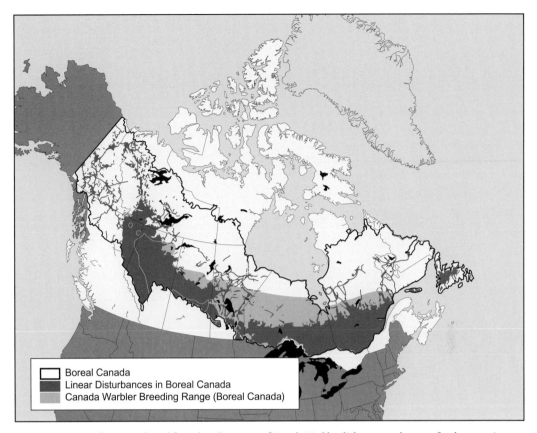

Figure 1.2. Overlap of Canadian boreal forest breeding range of Canada Warbler (light gray) and extent of anthropogenic disturbance (dark gray).

America's boreal forest. Over 45 million hectares of new protected areas have been established in Canada's boreal forest region since 2000 and 30 million hectares of forest tenures have been certified through the Forest Stewardship Council (S. Kallick, pers. comm.). The governments of Ontario and Quebec have pledged to establish another 80 million hectares of new protected areas and the federal government has continued to move forward on establishing new protected areas in the Northwest Territories that are co-managed with indigenous governments (Reid 2010). Many indigenous communities of Canada's boreal forest region have developed land-use plans that call for ambitious protected areas goals. For example, the Dehcho Land Use plan in the southern Northwest Territories calls for protecting at least 50% or nearly 11 million hectares of their lands (Charlwood and Wells 2007). In May of 2010, the Forest Products Association of Canada (FPAC) announced an agreement that it had reached with leading environmental organizations to halt logging on 29 million hectares of forestry tenures held by FPAC member companies in especially sensitive habitats while a conservation plan is developed for the total 72 million acres of forestry tenures that they control (Kallick 2010, Pew Environment Group 2010).

LITERATURE CITED

Anielski, M., and S. Wilson. 2006. Counting Canada's natural capital: assessing the real value of Canada's boreal ecosystems. Canadian Boreal Initiative and Pembina Institute, Ottawa, ON.

Bradshaw, C. J. A., I. G. Warkentin, and N. S. Sodhi. 2009. Urgent preservation of boreal carbon stocks and biodiversity. Trends in Ecology and Evolution 24:541–548.

Bryant, D., D. Nielsen, and L. Tangley. 1997. The last frontier forests. Ecosystems and economies on the edge. World Resources Institute, Washington DC.

Canadian Association of Petroleum Producers. 2005. CAPP releases 2004 petroleum reserves estimate. <http://www.capp.ca> (1 March 2006).

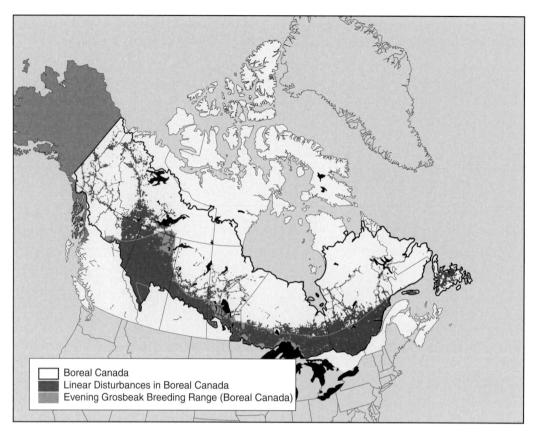

Figure 1.3. Overlap of Canadian boreal forest breeding range of Evening Grosbeak (light gray) and extent of anthropogenic disturbance (dark gray).

Canadian Boreal Initiative. 2005. The Boreal in the balance: securing the future of Canada's boreal region. Canadian Boreal Initiative, Ottawa, ON.

Canadian Council of Forest Ministers. 2008. Compendium of canadian forestry statistics. <http://nfdp.ccfm.org> (12 June 2010).

Cardillo, M., G. M. Mace, J. L. Gittleman, and A. Purvis. 2006. Latent extinction risk and the future battlegrounds of mammal conservation. Proceedings of the National Academy of Sciences 103:4157–4161.

Carlson, M., J. Chen, S. Elgie, C. Henschel, A. Montenegro, N. Roulet, N. Scott, C. Tarnocai, and J. Wells. 2010. Maintaining the role of Canada's forests and peatlands in climate regulation. The Forestry Chronicle 86:434–443.

Carlson, M., J. Wells, and D. Roberts. 2009. The carbon the world forgot: conserving the capacity of Canada's boreal forest region to mitigate and adapt to climate change. Boreal Songbird Initiative and Canadian Boreal Initiative, Seattle, WA and Ottawa, ON.

Charlwood, J., and J. Wells. 2007. Big conservation for big places: Dehcho First Nation raises the bar for boreal conservation. The All-Bird Bulletin, April.

Gauthier, J., and Y. Aubry. 1996. The breeding birds of Quebec. Province of Quebec Society for the Protection of Birds and Canadian Wildlife Service, Environment Canada, Quebec Region, Montreal, QC.

Hummel, M., and J. C. Ray. 2008. Caribou and the north: a shared future. Dundurn Press, Toronto, ON.

Kallick, S. 2010. World-class forest protection. Huffington Post. <http://www.huffingtonpost.com/steven-kallick/world-class-forest-protec_b_587798.html>.

Lee, P. D. Aksenov, L. Laestadius, R. Nogueron, and W. Smith. 2006. Canada's large intact forest landsapes. Global Forest Watch Canada, Edmonton, AB.

Pew Environment Group. 2010. Canadian forest industry and environmental groups sign world's largest conservation agreement applying to area twice the size of Germany. <http://www.pewtrusts.org/news_room_detail.aspx?id=58950>.

Reid, F. 2010. Progress and peril in the boreal forest. Ducks Unlimited Magazine, Spring. Ducks Unlimited, Memphis, TN.

Ricketts, T. H., E. Dinerstein, D. M. Olson, C. J. Loucks, W. Eichbaum, D. Dellasal, K. Kavanagh, P. Hedao, P. T. Hurley, K. M. Carney, R. Abell, and

S. Walters. 1999. Terrestrial ecoregions of North America: a conservation assessment. Island Press, Washington, DC.

Schindler, D., and P. Lee. 2010. Comprehensive conservation planning to protect biodiversity and ecosystem services in Canadian boreal regions under a warming climate and increasing exploitation. Biological Conservation 143:1571–1586.

Stanojevic, Z., P. Lee, and J. D. Gysbers. 2006a. Recent anthropogenic changes within the Boreal Plains ecozone of Saskatchewan and Manitoba: interim report. Global Forest Watch Canada, Edmonton, AB.

Stanojevic, Z., P. Lee, and J. D. Gysbers. 2006b. Recent anthropogenic changes within the Northern Boreal, Southern Taiga, and Hudson Plains ecozones of Québec. Global Forest Watch Canada, Edmonton, AB.

Tarnocai, C., J. G. Canadell, E. A. G. Schuur, P. Kuhry, G. Mazhitova, and S. Zimov. 2009. Soil organic carbon pools in the northern circumpolar permafrost region. Global Biogeochemical Cycles 23, GB2023, doi:10.1029/2008GB003327.

Timoney, K. and P. Lee. 2009. Does the Alberta tar sands industry pollute? The scientific evidence. Open Conservation Biology Journal, 3:65–81.

Wells, J., S. Casey-Lefkowitz, G. Chavarria, S. Dyer. 2008. Danger in the nursery: impact on birds of tar sands oil development in Canada's boreal forest. Boreal Songbird Initiative, Pembina Institute, and Natural Resources Defense Council, Seattle, WA, Calgary, AB, and Washington, DC.

Wells, J., D. Roberts, R. Cheng, P. Lee, and M. Darveau. 2011. A forest of blue—Canada's boreal forest: the world's waterkeeper. International Boreal Conservation Campaign and Canadian Boreal Initiative, Seattle, WA, and Ottawa, ON.

Wells, J. V. 2007. Birder's conservation handbook: 100 North American birds at risk. Princeton University Press, Princeton, NJ.

Wilcove, D. S. 2008. No way home: the decline of the world's great animal migrations. Island Press, Washington, DC.

CHAPTER TWO

Global Role for Sustaining Bird Populations

Jeffrey V. Wells and Peter J. Blancher

Abstract. Spanning 5.9 million square km, the boreal region of North America stretches from Alaska to Newfoundland, and represents 25% of the earth's remaining intact forests. As one of the largest wilderness areas left on the globe, the boreal forest hosts a diverse and unique avian assemblage that is a valuable component of global biodiversity. We endeavor to quantify its importance to the maintenance of North American bird populations. We used a simple modeling approach to develop an assessment of the conservation value of North America's boreal region for birds and provide one of the first attempts to quantitatively describe the stewardship responsibility of a global-scale ecosystem. Results illustrate that the boreal region is critical to the well-being of many bird species. We found that nearly half (325) of all regularly occurring North American bird species occur regularly within the boreal region and that over 300 species regularly breed there. In at least 96 species, 50% or more of their entire breeding population was estimated to occur within the boreal region. The boreal region of North America represents a unique global conservation asset for birds and other forms of biodiversity that should be protected.

Key Words: Alaska, boreal forest, biodiversity, bird conservation, Canada, conservation assessment, conservation value.

lobal-level conservation assessments have historically focused on measures of species diversity or rarity in the form of endemism or endangerment. Under such schemes, areas with high diversity and/or high numbers of endemic or endangered species have higher conservation value (Margules and Pressey 2000, Karieva and Marvier 2003, Cardillo et al. 2006, Ceballos and Ehrlich 2006, Lamoreux et al. 2006). Global-scale assessments that have used diversity and rarity-based approaches generally find that, as expected, tropical and subtropical regions of the world have the highest diversity and highest numbers of endemic and endangered species (Myers et al. 2000, Hoekstra et al. 2005, Wilson et al. 2006). In fact, diversity and rarity are but two of many conservation values that can be considered. For example, ecosystem function, ecosystem services, intactness of habitat, and species abundance have begun to be recognized as equally important factors to consider in developing conservation priorities (Costanza et al. 1997, Ricketts et al. 1999, Karieva and Marvier 2003, Mittermeier et al. 2003, Ceballos

Wells, J. V., and P. J. Blancher. 2011. Global role for sustaining bird populations. Pp. 7–22 *in* J. V. Wells (editor). Boreal birds of North America: a hemispheric view of their conservation links and significance. Studies in Avian Biology (no. 41), University of California Press, Berkeley, CA.

Figure 2.1. Bird conservation regions in North America's boreal forest region: northwest interior forest (region 4); boreal taiga plains (region 6); taiga shield and Hudson plains (region 7); boreal softwood shield (region 8). Adapted from Blancher and Wells (2005).

and Ehrlich 2006, Chan et al. 2006, Leroux and Schmiegelow 2007).

Conservation assessments based on diversity and rarity have been vastly important in raising awareness of the need for urgent conservation action to prevent the imminent extinction of many species and the loss of the last remnants of some of earth's most species-rich ecological communities. One consequence of such prioritization schemes, however, is that the conservation values and priorities of regions ranked relatively low in conservation priority are unclear. For example, a comparison of the bird species diversity of Ontario with that of Peru—two jurisdictions of roughly similar size—shows that Peru has vastly greater species diversity (1,800 spp versus 500 spp). Not surprisingly, though, the Ontario government will not spend the bulk of its taxpayer monies on biodiversity protection in Peru despite the fact that Peru has more species. That is because most national and provincial or state governments and nonprofit organizations are given mandates to protect and manage species within their geographic boundaries. In this example, the Ontario government, rather than comparing its species diversity to that of another country, would be better served to understand and enumerate the conservation values that make it globally unique.

One measure of conservation value that has only recently been given explicit consideration in conservation planning is that of abundance. Through its Important Bird Areas program, BirdLife International was one of the first organizations to

explicitly consider both abundance and rarity in conservation decision-making (Wells 1998, Heath and Evans 2000, Fishpool and Evans 2001, Chipley et al. 2003, Wells et al. 2005). More recently, the Partners in Flight coalition and its members have highlighted the concept that some regions have a high stewardship responsibility for maintaining species that are still abundant (Rosenberg and Wells 1995, 2000, 2005; Wells and Rosenberg 1999; Rich et al. 2004).

In this paper, we provide an assessment of the value of North America's boreal region in supporting and maintaining continental bird populations. We consider the taxonomic diversity supported by the boreal region at the family and species level, the total numbers of birds of each species that breed in the boreal region, and the use of the boreal region by birds during migration.

METHODS

The boreal forest of North America, stretching from Alaska across 6,000 km to Newfoundland, is, at 5.9 million square km, the largest wilderness left in North America and represents 25% of the world's remaining intact forests (Canadian Boreal Institute 2005). Although there are many delineations of the boreal zone that differ slightly in the mapping of the southern and/or northern border, for these analyses we defined the boreal region as the area within Bird Conservation Regions 4, 6, 7, and 8 (Fig. 2.1; U.S. NABCI Committee 2000).

Distribution maps of all North American birds (Ridgely et al. 2003) were overlaid with boundaries of Bird Conservation Regions (BCRs) and the proportion of the breeding, migration, and wintering range of each species that occurred within the boreal region was calculated using ArcINFO. An implicit assumption in the use of distribution maps is that a region's importance to a species is strongly related to the proportion of that species' range in the region. This assumption is reasonable for most species, but may break down for species with highly clumped distributions, such as breeding seabirds and other colonial waterbirds. For these species, use of colony counts if available across a species range would provide a more accurate assessment of relative importance of the boreal region. North American Breeding Bird Survey (BBS) data from 1990 to 1999 were used to provide an alternative measure of the proportion of breeding birds of each species in the boreal region using methods outlined in Rosenberg and Wells (1995, 2000). We calculated a mean relative abundance for each species within portions of each BCR that occurred within a province or state. A correction factor was applied to each mean value to account for differences in detectability and time of day and to correct for male-biased detection of singing birds, as described in Rich et al. (2004). Each adjusted mean abundance value was divided by the estimated area of detectability to obtain an estimate of density, and this value was multiplied by the area of each provincial/state BCR polygon to obtain an estimated total abundance for that polygon. Total abundance for each species within each polygon was summed across all polygons where the species occurred to get a North American total population size. Abundance estimates from the four boreal BCRs were summed and the percent of total North American population size was calculated for each species.

BBS relative abundance estimates in the boreal region were based on a reasonably high number of individual survey routes (265). However, the distribution of routes is biased toward the southern parts of the boreal. Routes were stratified by province/state/territory and BCR to minimize effects of this bias, but low sample size in the northern parts of the boreal results in low precision in estimates of bird numbers for many species. Some species, particularly many nonlandbirds, were not sampled well by BBS surveys. For this reason, estimates of population sizes of shorebirds, waterfowl, and waterbirds relied on continental estimates provided in continental plans (Donaldson et al. 2000 and Brown et al. 2001 for shorebirds, Kushlan et al. 2002 and Milko et al. 2003 for waterbirds, and NAWMP 2004 for waterfowl). Continental estimates were multiplied by proportions of range or proportions of BBS population in the boreal to give an approximate estimate of breeding population size in the boreal. For waterfowl in particular, a more accurate estimate of the proportion of continental populations that occur in the boreal should be possible with the use of various waterfowl survey data sets, not treated in this paper (but see Chapter 3).

Proportions of western hemisphere population for all birds were based on a combination of BBS proportions within the BBS survey area and proportion of breeding range elsewhere in the hemisphere.

RESULTS

Taxonomic Diversity

Nearly 400 species (399, or 57% of regularly occurring birds of the U.S. and Canada) are known to occur within some portion of the boreal forest region of Alaska and Canada. Excluding species that are exclusively marine or coastal or that occur in less than 1% of the boreal region, there remain 325 species (47%) (Appendix 2.1—see this appendix for all scientific names of species mentioned in text). A total of 304 species (43%) breed in the forests, thickets, and wetlands of the boreal forest. The remaining 21 species occur as migrants or winterers within the region.

At least 47 families of birds are represented, making up 67% of all bird families that regularly occur in the U.S. and Canada. Certain families have an especially high representation in the boreal. Thirty-five of 44 waterfowl species (80%) that nest in the U.S. or Canada are boreal forest breeders, at least in part. Fifty-three percent (27 of 51) of warblers, 63% of finches, and 93% (13 of 14) thrushes that nest in the U.S. and Canada are boreal breeders.

Population Estimates

The number of birds breeding in North America's boreal forest is estimated at between 1.65 and 3 billion (Table 2.1). Landbirds are by far the

TABLE 2.1
Estimated number of birds breeding in North America's boreal forest by bird group

Bird Group	Estimated Breeding Population	% of U.S./ Canadian Population
Landbirds	1,600,000,000	30%
Shorebirds	7,000,000	30%
Waterbirds	14,000,000	N/A
Waterfowl	26,000,000	38%

most numerous, making up 97% of all individuals breeding in the boreal. An estimated 38% (26 million) of all the waterfowl of Canada and the U.S. breed in the boreal. Approximately 30% of all shorebirds (7 million) and 30% of all landbirds (1–3 billion) that breed in the U.S. and Canada occur within the boreal.

Species Abundance

A total of 276 species have 5% or more of their breeding range within the boreal forest. Ninety-six species representing 14% of the total U.S./ Canadian avifauna have 50% or more of their estimated total breeding population within the boreal (Box 2.1). Another 55 species have between 25% and 50% of their breeding population within the boreal (Box 2.2). A wide variety of birds are represented among these boreal birds, including several species from each major bird group: waterfowl, waterbirds, shorebirds, and landbirds. More than 80% of the populations of 35 species are found in the boreal forest (Box 2.3), including Palm Warbler (>90% of population in boreal), Short-billed Dowitcher (>90%), Northern Shrike (>90%), Bonaparte's Gull (>90%), Spruce Grouse (>90%), Red-necked Grebe (>90%), Gray Jay (80–90%), Bufflehead (80–90%), White-winged Scoter (80–90%), Rusty Blackbird (80–90%), and Great Gray Owl (80–90%).

Non-boreal Breeding Migrants

Virtually all species of boreal nesting birds also make use of parts of the boreal during migration. We categorized 29 species as relying on the boreal more for migratory stop-over habitat than for breeding or wintering (Box 2.4), and some of these do not breed anywhere in the boreal. For example, the White-rumped Sandpiper does not breed in the boreal region but makes extensive use of wetlands within the boreal during its fall and spring migration. Other shorebirds, like the Pectoral Sandpiper, that have insignificant portions of the breeding range in the boreal zone are also highly reliant on boreal forest wetlands during migration. Waterfowl like Greater White-fronted Goose, Snow Goose, Cackling Goose, Tundra Swan, and Greater Scaup also regularly migrate through a large part of the boreal region. Not quantified here is use of the boreal region by "molt-migrants," birds that migrate north into the boreal forest after breeding to undergo molt, a practice common among many waterfowl species.

DISCUSSION

Our assessment provides a more complete view of the conservation importance of a global-scale ecoregion that is outside the traditionally identified world biodiversity hotpots. Our analyses show that the North American boreal region supports over 50% of the North American populations of at least 96 species. Consequently, the boreal region has large stewardship responsibility to these species because land-use decisions within the region have the potential to affect the bulk of the species total populations and therefore will have a large effect on species persistence and long-term extinction probability. For the 35 species we identified as having 80% or more of their North American population occurring within the boreal region, the issue of how their habitat is managed is even more critical. At least eight of these 35 species have shown significant declines (Brown et al. 2001, Blancher 2003, NAWMP 2004, Rich et al.

BOX 2.1

Species with 50% or more of estimated Western Hemisphere breeding population in North America's boreal forest (96 Species)

Trumpeter Swan
American Wigeon
American Black Duck
Green-winged Teal
Ring-necked Duck
Greater Scaup
Lesser Scaup
Surf Scoter
White-winged Scoter
Black Scoter
Bufflehead
Common Goldeneye
Barrow's Goldeneye
Hooded Merganser
Common Merganser
Ruffed Grouse
Spruce Grouse
White-tailed Ptarmigan
Pacific Loon
Common Loon
Horned Grebe
Red-necked Grebe
Merlin
Yellow Rail
Sora
Whooping Crane
Semipalmated Plover
Greater Yellowlegs
Lesser Yellowlegs
Solitary Sandpiper
Wandering Tattler
Spotted Sandpiper
Whimbrel
Hudsonian Godwit
Surfbird
Least Sandpiper
Short-billed Dowitcher
Wilson's Snipe
Red-necked Phalarope
Little Gull
Bonaparte's Gull
Mew Gull
Herring Gull
Common Tern
Arctic Tern
Northern Hawk Owl
Great Gray Owl
Boreal Owl

Yellow-bellied Sapsucker
American Three-toed Woodpecker
Black-backed Woodpecker
Olive-sided Flycatcher
Yellow-bellied Flycatcher
Alder Flycatcher
Least Flycatcher
Northern Shrike
Blue-headed Vireo
Philadelphia Vireo
Gray Jay
Boreal Chickadee
Gray-headed Chickadee
Ruby-crowned Kinglet
Gray-cheeked Thrush
Swainson's Thrush
Hermit Thrush
Bohemian Waxwing
Tennessee Warbler
Orange-crowned Warbler
Nashville Warbler
Magnolia Warbler
Cape May Warbler
Yellow-rumped Warbler
Black-throated Green Warbler
Blackburnian Warbler
Palm Warbler
Bay-breasted Warbler
Blackpoll Warbler
Black-and-white Warbler
Northern Waterthrush
Connecticut Warbler
Mourning Warbler
Wilson's Warbler
Canada Warbler
Clay-colored Sparrow
Le Conte's Sparrow
Fox Sparrow
Lincoln's Sparrow
Swamp Sparrow
White-throated Sparrow
White-crowned Sparrow
Golden-crowned Sparrow
Dark-eyed Junco
Rusty Blackbird
Gray-crowned Rosy-Finch
Pine Grosbeak
White-winged Crossbill

BOX 2.2

Species with 25–49% of estimated Western Hemisphere breeding population in North America's boreal forest (55 Species)

Greater White-fronted Goose
Canada Goose
Northern Shoveler
Northern Pintail
Common Eider
Long-tailed Duck
Red-breasted Merganser
Double-crested Cormorant
American Bittern
Osprey
Sharp-shinned Hawk
Northern Goshawk
Broad-winged Hawk
Sandhill Crane
American Golden-Plover
Semipalmated Sandpiper
Dunlin
Stilt Sandpiper
American Woodcock
Franklin's Gull
Black Tern
Long-eared Owl
Belted Kingfisher
Hairy Woodpecker
Northern Flicker
Western Wood-Pewee
Eastern Phoebe
Red-eyed Vireo

Black-billed Magpie
Tree Swallow
Bank Swallow
Black-capped Chickadee
Red-breasted Nuthatch
Winter Wren
Golden-crowned Kinglet
Arctic Warbler
Veery
American Robin
Varied Thrush
American Pipit
Cedar Waxwing
Yellow Warbler
Chestnut-sided Warbler
American Redstart
Ovenbird
American Tree Sparrow
Chipping Sparrow
Savannah Sparrow
Nelson's Sharp-tailed Sparrow
Smith's Longspur
Rose-breasted Grosbeak
Purple Finch
Common Redpoll
Pine Siskin
Evening Grosbeak

2004, Wells 2007), putting them at even greater risk from habitat loss, degradation, disturbance, and changes.

Another unintended consequence of the focus of conservation science on diversity, endemism, and endangerment is that these same conservation values are applied at scales much finer than the global level. For example, a U.S. state agency or even a local park or refuge may give greater value to species that are locally rare but globally common while ignoring a species that is locally common but globally rare (Wells et al. 2010).

While we have focused on abundance, there are other conservation values that can and should be considered (Jepson and Canney 2001), including intactness of habitat (Bryant et al. 1997, Aplet et al. 2000, Aksenov et al. 2002, Lee et al. 2003, Mittermeier et al. 2003), integrity of natural system cycles like migration and irruptive movements (Wilcove 2008), predator–prey dynamics, adaptation to insect outbreaks and fire-induced habitat change, resilience to and capacity for buffering against climate change (Root and Schneider 2002), maintenance of ecosystem services and natural capital (Anielski and Wilson 2006, Chan et al. 2006), and traditional aboriginal values (Stevenson and Webb 2004).

The importance of protecting areas with globally unique levels of species diversity, endemism, and endangerment has been well established in

BOX 2.3

Species with more than 80% of estimated Western Hemisphere breeding population in North America's boreal forest (35 Species)

Surf Scoter
White-winged Scoter
Black Scoter
Bufflehead
Common Goldeneye
Spruce Grouse
Red-necked Grebe
Whooping Crane
Lesser Yellowlegs
Solitary Sandpiper
Surfbird
Short-billed Dowitcher
Bonaparte's Gull
Herring Gull
Great Gray Owl
American Three-toed Woodpecker
Black-backed Woodpecker
Yellow-bellied Flycatcher

Alder Flycatcher
Northern Shrike
Philadelphia Vireo
Gray Jay
Boreal Chickadee
Bohemian Waxwing
Tennessee Warbler
Cape May Warbler
Palm Warbler
Blackpoll Warbler
Connecticut Warbler
Lincoln's Sparrow
White-throated Sparrow
Dark-eyed Junco
Rusty Blackbird
Pine Grosbeak
White-winged Crossbill

BOX 2.4

Species in which the area of North America's boreal forest occupied during migration exceeds the area occupied during breeding or wintering (29 Species)

Greater White-fronted Goose
Snow Goose
Ross's Goose
Brant
Cackling Goose
Tundra Swan
Gadwall
Rough-legged Hawk
Whooping Crane
Black-bellied Plover
American Golden-Plover
Pacific Golden-Plover
Semipalmated Plover
Hudsonian Godwit
Red Knot

Sanderling
Semipalmated Sandpiper
White-rumped Sandpiper
Baird's Sandpiper
Pectoral Sandpiper
Dunlin
Stilt Sandpiper
Buff-breasted Sandpiper
Long-billed Dowitcher
American Pipit
Harris's Sparrow
Lapland Longspur
Smith's Longspur
Snow Bunting

conservation science, as has the identification of such areas at the global scale (Stattersfield et al. 1998, Myers et al. 2000, Olson and Dinerstein 2002). Other conservation values also have a role to play in schemes that prioritize regions and allocate conservation efforts (Jepson and Canney 2001). We focused on the value of abundance and provided an assessment of the global significance of that value for North America's boreal forest region, one of the earth's largest and most intact terrestrial ecoregions. Our assessment should not be considered to conflict with assessments that show tropical regions are of crucial importance for maintaining high species diversity and endemism and that such regions are often under extreme threat from human development and industrial pressures. Tropical regions deserve major support from globally focused organizations as well as from the public and from the governments and organizations working directly within those regions. But the rest of the world's ecoregions are clearly not without their own important conservation values. Conservation science needs to use caution in ensuring that the application of space-based prioritization using conservation values like diversity, endemism, and endangerment does not cause governments and organizations that work within a country or region to overlook the conservation values of most significance in their own region.

ACKNOWLEDGEMENTS

This report was modeled after a previous report on the importance of Canada's boreal forest to landbirds in the Western Hemisphere (Blancher 2003). However, analyses reported here were updated to incorporate more recent data and expanded to include non-landbirds and the full extent of the North American boreal, including much of Alaska.

Species range information came largely from Ridgely et al. (2003) shape files. These data were provided by NatureServe in collaboration with Robert Ridgely, James Zook, The Nature Conservancy's Migratory Bird Program, Conservation International's Center for Applied Biodiversity Science, World Wildlife Fund–U.S., and Environment Canada–WILDSPACE. Thanks especially to Andrew Couturier, Bird Studies Canada, who overlaid these shape files onto jurisdictional maps, a BCR layer, and latitude/longitude degree blocks, thus enabling us to calculate proportions of range for each species. Thanks to Environment Canada, Ontario Region, for providing the BCR shape file used for analyses in this report.

Breeding Bird Survey data were obtained from the useful U.S. Geological Survey web pages devoted to this survey. The methods of estimating population size from BBS data were developed in conjunction with Ken Rosenberg (Cornell Lab of Ornithology) and with input from the Partners in Flight Science Committee during development of the PIF North American Landbird Conservation Plan (Rich et al. 2004). We would also like to thank the thousands of volunteers who collected the data used here. We thank Bruce Robertson and Iain Steinhouse for helpful comments on the manuscript.

LITERATURE CITED

Aksenov, D., D. Dobrynin, M. Dubinin, A. Egorov, A. Isaev, M. Karpachevskiy, L. Laestadius, P. Potapov, A. Purekhovskiy, S. Turubanova, and A. Yaroshenko. 2002. Atlas of Russia's intact forest landscapes. Global Forest Watch Russia, Moscow.

Anielski, M., and S. Wilson. 2006. Counting Canada's natural capital: assessing the real value of Canada's boreal ecosystems. Canadian Boreal Initiative and Pembina Institute, Ottawa, ON.

Aplet, G., J. Thomson, and M. Wilbert. 2000. Indicators of wildness: using attributes of the land to assess the context of wilderness. In D. N. Cole and S. F. McCool (editors), Proceedings: Wilderness science in a time of change. Proceedings RMRS-P-15. USDA Forest Service, Rocky Mountain Research Station, Ogden, UT.

Blancher, P. 2003. Importance of Canada's boreal forest to landbirds. Canadian Boreal Initiative and Boreal Songbird Initiative, Ottawa, ON, and Seattle WA.

Blancher, P., and J. V. Wells. 2005. The boreal forest region: North America's bird nursery. Canadian Boreal Initiative and Boreal Songbird Initiative, Ottawa, ON, and Seattle, WA.

Brown, S., C. Hickey, B. Harrington, and R. Gill (editors). 2001. United States shorebird conservation plan. 2nd ed. Manomet Center for Conservation Sciences, Manomet, MA.

Bryant, D., Nielsen, D. and Tangley, L. 1997. The last frontier forests: ecosystems and economies on the edge. World Resources Institute, Washington, DC.

Canadian Boreal Initiative. 2005. The boreal in the balance: securing the future of Canada's boreal region. Canadian Boreal Initiative, Ottawa, ON.

Cardillo, M., G. M. Mace, J. L. Gittleman, and A. Purvis. 2006. Latent extinction risk and the future battlegrounds of mammal conservation. Proceedings of the National Academy of Sciences 103:4157–4161.

Ceballos, G., and P. R. Ehrlich. 2006. Global mammal distributions, biodiversity hotspots, and conservation. Proceedings of the National Academy of Sciences 103:19374–19379.

Chan, K. C., M. R. Shaw, D. R. cameron, E. C. Underwood, and G. C. Daily. 2006. Conservation planning for ecosystem services. PLoS Biology 4:2138–2152.

Chipley, R. M., G. H. Fenwick, M.J . Parr, and D. N. Pashley. 2003. The American Bird Conservancy guide to the 500 most important bird areas in the United States. Random House, New York, NY.

Costanza, R., R., Darge, R. Degroot, S. Farber, M. Grasso, B. Hannon, K. Limburg, S. Naeem, R. V. O'Neill, J. Paruelo, R. G. Raskin, P. Sutton, and M. Vandenbelt. 1997. The value of the world's ecosystem services and natural capital. Nature 387: 253–260.

Donaldson, G. M., C. Hyslop, R. I. G. Morrison, H. L. Dickson, and I. Davidson (editors). 2000. Canadian shorebird conservation plan. Canadian Wildlife Service, Hull, QC.

Fishpool, L. D. C., and M. I. Evans. 2001. Important bird areas in Africa and associated islands: priority sites for conservation. BirdLife International, Cambridge, UK.

Heath, M. F., and M. I. Evans (editors). 2000. Important bird areas in Europe: priority sites for conservation. BirdLife International, Cambridge, UK.

Hoekstra, J. M., T. M. Boucher, T. H. Ricketts, and C. Roberts. 2005. Confronting a biome crisis: global disparities of habitat loss and protection. Ecology Letters 8:23–29.

Jepson, P., and S. Canney. 2001. Biodiversity hotspots: hot for what? Global Ecology and Biogeography 10:225–227.

Karieva, P., and M. Marvier. 2003. Conserving biodiversity coldspots. American Scientist 91:344–351.

Kushlan, J. A., M. J. Steinkamp, K. C. Parsons, J. Capp, M. A. Cruz, M. Coulter, I. Davidson, L. Dickson, N. Edelson, R. Elliot, R. M. Erwin, S. Hatch, S. Kress, R. Milko, S. Miller, K. Mills, R. Paul, R. Phillips, J. E. Saliva, B. Sydeman, J. Trapp, J. Wheeler, and K. Wohl. 2002. Waterbird conservation for the Americas: The North American waterbird conservation plan, version 1. Waterbird Conservation for the Americas, Washington, DC.

Lamoreux. J. F., J. C. Morrison, T. H, Ricketts, D. M. Olson, E. Dinerstein, M. W. McKnight, and H. H. Shugart. 2006. Global tests of biodiversity concordance and the importance of endemism. Nature 440:212–214.

Lee, P., D. Aksenov, L. Laestadius, R. Nogueron, and W. Smith. 2006. Canada's large intact forest landsapes. Global Forest Watch Canada, Edmonton, AB.

Lee, P., D. Aksenov, L. Laestadius, R. Noguevon, and W. Smith 2003. Canada's large intact forest landscapes. Global Forest Watch Canada, Edmonton, AB, and World Resources Institute, Washington, DC.

Leroux, S. J., and F. K. A. Schmiegelow. 2007. Biodiversity concordance and the importance of endemism. Conservation Biology 21:266–268.

Margules, C. R., and R. L. Pressey. 2000. Systematic conservation planning. Nature 405:243–253.

Milko, R., L. Dickson, R. Elliot, and G. Donaldson. 2003. Wings over water: Canada's waterbird conservation plan. Canadian Wildlife Service, Ottawa, ON.

Mittermeier, R. A., C. G. Mittermeier, T. M. Brooks, J. D. Pilgrim, W. R. Konstant, G. A. B. Da Fonseca, and C. Kormos. 2003. Wilderness and biodiversity conservation. Proceedings of the National Academy of Sciences 100:10309–10313.

Myers, N., R. A. Mittermeir, C. G. Mittermeir, G. A. B. da Fonseca, and J. Kent. 2000. Biodiversity hotspot for conservation priorities. Nature 403:853–858.

NAWMP. 2004. 2004 North American waterfowl management plan: strengthening the biological foundation. U.S. Fish and Wildlife Service, Arlington, VA; Direccion General de Vida Silvestre, Mexico, DF; and Canadian Wildlife Service, Gatineau, QC.

Olson, D. M., and E. Dinerstein. 2002. The global 200: priority ecoregions for global conservation. Annals of the Missouri Botanical Gardens. 89:199–224.

Rich, T. D., C. J. Beardmore, H. Berlanga, P. J. Blancher, M. S. W. Bradstreet, G. S. Butcher, D. W. Demarest, E. H. Dunn, W. C. Hunter, E. E. Inigo-Elias, J. A. Kennedy, A. M. Martell, A. O. Panjabi, D. N. Pashley, K. V. Rosenberg, C. M. Rustay, J. S. Wendt and T. C. Will. 2004. Partners in Flight North American landbird conservation plan. Cornell Lab of Ornithology, Ithaca, NY.

Ricketts, T. H., E. Dinerstein, D. M. Olson, C. J. Loucks, et al. 1999. Terrestrial ecoregions of North America: a conservation assessment. Island Press, Washington, DC.

Ridgely, R. S., T. F. Allnutt, T. Brooks, D. K. McNicol, D. W. Mehlman, B. E. Young, and J. R. Zook. 2003. Digital distribution maps of the birds of the western hemisphere, version 1.0. NatureServe, Arlington, VA.

Root, T. L., and S. H. Schneider. 2002. Climate change: overview and implications for wildlife. Pp. 1-56 in S. H. Schneider and T .L. Root (editors), Wildlife responses to climate change: North American case studies. Island Press, Washington, DC.

Rosenberg, K. V., and J. V. Wells. 1995. Importance of geographic areas to neotropical migrants in the Northeast. Final report to U.S. Fish and Wildlife Service, Region 5, Hadley, MA.

Rosenberg, K. V., and J. V. Wells. 2000. Global perspectives on neotropical migratory bird conservation in the Northeast: long-term responsibility versus immediate concern. Pp. 32–43 in R. Bonney, D. N. Pashley, R. J. Cooper, and L. Niles (editors), Strategies for bird conservation: the Partners In Flight planning process.

Proceedings of the 3rd Partners in Flight Workshop, 1–5 October 1995, Cape May, NJ. Proceedings RMRS-P-16. U.S. Department of Agriculture, Forest Service, Rocky Mountain Research Station, Ogden, UT.

Rosenberg, K. V., and J. V. Wells. 2005. Conservation priorities for terrestrial birds in the northeastern United States. Pp. 236–253 *in* C. J. Ralph and T. D. Rich (editors), Bird conservation implementation and integration in the Americas. Proceedings of the Third International Partners in Flight Conference, 20–24 March 2002, Asilomar, CA, Vol. 1. General Technical Report PSW-GTR-191. USDA Forest Service, Albany, CA.

Stattersfield, A. J., M. J. Crosby, A. J. Long, and D. C. Wege. 1998. Endemic bird areas of the world: priorities for biodiversity conservation. BirdLife International, Cambridge, UK.

Stevenson, M. G., and J. Webb. 2004. First Nations: measures and monitors of boreal forest biodiversity. Ecological Bulletins 51:83–92.

U.S. NABCI Committee. 2000. North American Bird Conservation Initiative. Bird Conservation Region Descriptions. A Supplement to the North American Bird Conservation Initiative Bird Conservation Regions Map. U.S. Fish and Wildlife Service, Arlington, VA.

Wells, J. V. 1998. Important bird areas in New York State. National Audubon Society, Albany, NY.

Wells, J. V. 2007. Birder's conservation handbook: 100 North American birds at risk. Princeton University Press, Princeton, NJ.

Wells, J. V., B. Robertson, K. V. Rosenberg , and D. W. Mehlman. 2010. Global versus local conservation focus of U.S. state agency endangered bird species lists. PLoS ONE 5(1). e8608. doi:10.1371/journal.pone.0008608.

Wells, J. V., and K. V. Rosenberg. 1999. Grassland bird conservation in northeastern North America. Studies in Avian Biology 19:72–80.

Wells, J. V., D. K. Niven, and J. Cecil. 2005. The Important Bird Areas program in the United States: building a network of sites for conservation, state by state. Pp. 1265–1269 *in* C. J. Ralph and T. D. Rich (editors), Bird conservation implementation and integration in the Americas. Proceedings of the Third International Partners in Flight Conference, 20–24 March 2002, Asilomar, CA, Vol. 2. General Technical Report PSW-GTR-191. USDA Forest Service, Albany, CA.

Wilcove, D. S. 2008. No way home: the decline of the world's great animal migrations. Island Press, Washington, DC.

Wilson, K. A., M. F. McBride, M. Bode, and H. P. Possingham. 2006. Prioritizing global conservation efforts. Nature 440:337–340.

Bird species that regularly occur in North America's boreal forest during breeding, migration, or wintering seasons (325 Species)

Greater White-fronted Goose (*Anser albifrons*)

Snow Goose (*Chen caerulescens*)

Ross's Goose (*Chen rossii*)

Brant (*Branta bernicla*)

Cackling Goose (*Branta hutchinsii*)

Canada Goose (*Branta canadensis*)

Trumpeter Swan *(Cygnus buccinator)*

Tundra Swan (*Cygnus columbianus*)

Wood Duck (*Aix sponsa*)

Gadwall (*Anas strepera*)

American Wigeon (*Anas americana*)

American Black Duck (*Anas rubripes*)

Mallard (*Anas platyrhynchos*)

Blue-winged Teal (*Anas discors*)

Northern Shoveler (*Anas clypeata*)

Northern Pintail (*Anas acuta*)

Green-winged Teal (*Anas crecca*)

Canvasback (*Aythya valisineria*)

Redhead (*Aythya americana*)

Ring-necked Duck (*Aythya collaris*)

Greater Scaup (*Aythya marila*)

Lesser Scaup (*Aythya affinis*)

Common Eider (*Somateria mollissima*)

Harlequin Duck (*Histrionicus histrionicus*)

Surf Scoter (*Melanitta perspicillata*)

White-winged Scoter (*Melanitta fusca*)

Black Scoter (*Melanitta americana*)

Long-tailed Duck (*Clangula hyemalis*)

Bufflehead (*Bucephala albeola*)

Common Goldeneye (*Bucephala clangula*)

Barrow's Goldeneye (*Bucephala islandica*)

Hooded Merganser (*Lophodytes cucullatus*)

Common Merganser (*Mergus merganser*)

Red-breasted Merganser (*Mergus serrator*)

Ruddy Duck (*Oxyura jamaicensis*)

Gray Partridge (*Perdix perdix*)

Ring-necked Pheasant (*Phasianus colchicus*)

Ruffed Grouse (*Bonasa umbellus*)

Spruce Grouse (*Falcipennis canadensis*)

Willow Ptarmigan (*Lagopus lagopus*)

Rock Ptarmigan (*Lagopus muta*)

White-tailed Ptarmigan (*Lagopus leucura*)

Dusky Grouse (*Dendragapus obscurus*)

Sharp-tailed Grouse (*Tympanuchus phasianellus*)

Red-throated Loon (*Gavia stellata*)

Pacific Loon (*Gavia pacifica*)

Common Loon (*Gavia immer*)

Yellow-billed Loon (*Gavia adamsii*)

Pied-billed Grebe (*Podilymbus podiceps*)

Horned Grebe (*Podiceps auritus*)

Red-necked Grebe (*Podiceps grisegena*)

Eared Grebe (*Podiceps nigricollis*)

Western Grebe (*Aechmophorus occidentalis*)

American White Pelican (*Pelecanus erythrorhynchos*)

Double-crested Cormorant (*Phalacrocorax auritus*)

American Bittern (*Botaurus lentiginosus*)

Great Blue Heron (*Ardea herodias*)

Black-crowned Night-Heron (*Nycticorax nycticorax*)

Osprey (*Pandion haliaetus*)

Bald Eagle (*Haliaeetus leucocephalus*)

Northern Harrier (*Circus cyaneus*)

Sharp-shinned Hawk (*Accipiter striatus*)

Cooper's Hawk (*Accipiter cooperii*)

Northern Goshawk (*Accipiter gentilis*)

Red-shouldered Hawk (*Buteo lineatus*)

Broad-winged Hawk (*Buteo platypterus*)

Swainson's Hawk (*Buteo swainsoni*)

Red-tailed Hawk (*Buteo jamaicensis*)

Rough-legged Hawk (*Buteo lagopus*)

Golden Eagle (*Aquila chrysaetos*)

American Kestrel (*Falco sparverius*)

Merlin (*Falco columbarius*)

Gyrfalcon (*Falco rusticolus*)

Peregrine Falcon (*Falco peregrinus*)

Prairie Falcon (*Falco mexicanus*)

Yellow Rail (*Coturnicops noveboracensis*)

Virginia Rail (*Rallus limicola*)

Sora (*Porzana carolina*)

American Coot (*Fulica americana*)

Sandhill Crane (*Grus canadensis*)

Whooping Crane (*Grus americana*)

Black-bellied Plover (*Pluvialis squatarola*)

American Golden-Plover (*Pluvialis dominica*)

Pacific Golden-Plover (*Pluvialis fulva*)

Semipalmated Plover (*Charadrius semipalmatus*)

Piping Plover (*Charadrius melodus*)

Killdeer (*Charadrius vociferus*)

Eurasian Dotterel (*Charadrius morinellus*)

American Avocet (*Recurvirostra americana*)

Greater Yellowlegs (*Tringa melanoleuca*)

Lesser Yellowlegs (*Tringa flavipes*)

Solitary Sandpiper (*Tringa solitaria*)

Willet (*Tringa semipalmata*)

Wandering Tattler (*Tringa incana*)

Spotted Sandpiper (*Actitis macularius*)

Upland Sandpiper (*Bartramia longicauda*)

Whimbrel (*Numenius phaeopus*)

Bristle-thighed Curlew (*Numenius tahitiensis*)

Hudsonian Godwit (*Limosa haemastica*)

Bar-tailed Godwit (*Limosa lapponica*)

Marbled Godwit (*Limosa fedoa*)

Ruddy Turnstone (*Arenaria interpres*)

Black Turnstone (*Arenaria melanocephala*)

Surfbird (*Aphriza virgata*)

Red Knot (*Calidris canutus*)

Sanderling (*Calidris alba*)

Semipalmated Sandpiper (*Calidris pusilla*)

Western Sandpiper (*Calidris mauri*)

Least Sandpiper (*Calidris minutilla*)

White-rumped Sandpiper (*Calidris fuscicollis*)

Baird's Sandpiper (*Calidris bairdii*)

Pectoral Sandpiper (*Calidris melanotos*)

Purple Sandpiper (*Calidris maritima*)

Rock Sandpiper (*Calidris ptilocnemis*)

Dunlin (*Calidris alpina*)

Stilt Sandpiper (*Calidris himantopus*)

Buff-breasted Sandpiper (*Tryngites subruficollis*)

Short-billed Dowitcher (*Limnodromus griseus*)

Long-billed Dowitcher (*Limnodromus scolopaceus*)

Wilson's Snipe (*Gallinago delicata*)

American Woodcock (*Scolopax minor*)

Wilson's Phalarope (*Phalaropus tricolor*)

Red-necked Phalarope (*Phalaropus lobatus*)

Red Phalarope (*Phalaropus fulicarius*)

Pomarine Jaeger (*Stercorarius pomarinus*)

Parasitic Jaeger (*Stercorarius parasiticus*)

Long-tailed Jaeger (*Stercorarius longicaudus*)

Franklin's Gull (*Leucophaeus pipixcan*)

Little Gull (*Hydrocoloeus minutus*)

Black-headed Gull (*Chroicocephalus ridibundus*)

Bonaparte's Gull (*Chroicocephalus philadelphia*)

Mew Gull (*Larus canus*)

Ring-billed Gull (*Larus delawarensis*)

California Gull (*Larus californicus*)

Herring Gull (*Larus argentatus*)

Thayer's Gull (*Larus thayeri*)

Iceland Gull (*Larus glaucoides*)

Lesser Black-backed Gull (*Larus fuscus*)

Glaucous-winged Gull (*Larus glaucescens*)

Glaucous Gull (*Larus hyperboreus*)

Sabine's Gull (*Xema sabini*)

Ross's Gull (*Rhodostethia rosea*)

Caspian Tern (*Hydroprogne caspia*)

Common Tern (*Sterna hirundo*)

Arctic Tern (*Sterna paradisaea*)

Forster's Tern (*Sterna forsteri*)

Black Tern (*Chlidonias niger*)

Rock Pigeon (*Columba livia*)

Mourning Dove (*Zenaida macroura*)

Black-billed Cuckoo (*Coccyzus erythropthalmus*)

Great Horned Owl (*Bubo virginianus*)

Snowy Owl (*Bubo scandiacus*)

Northern Hawk Owl (*Surnia ulula*)

Northern Pygmy-Owl (*Glaucidium gnoma*)

Barred Owl (*Strix varia*)

Great Gray Owl (*Strix nebulosa*)

Long-eared Owl (*Asio otus*)

Short-eared Owl (*Asio flammeus*)

Boreal Owl (*Aegolius funereus*)

Northern Saw-whet Owl (*Aegolius acadicus*)

Common Nighthawk (*Chordeiles minor*)

Whip-poor-will (*Caprimulgus vociferus*)

Chimney Swift (*Chaetura pelagica*)

Ruby-throated Hummingbird (*Archilochus colubris*)

Rufous Hummingbird (*Selasphorus rufus*)

Belted Kingfisher (*Megaceryle alcyon*)

Lewis's Woodpecker (*Melanerpes lewis*)

Red-headed Woodpecker (*Melanerpes erythrocephalus*)

Yellow-bellied Sapsucker (*Sphyrapicus varius*)

Red-naped Sapsucker (*Sphyrapicus nuchalis*)

Red-breasted Sapsucker (*Sphyrapicus ruber*)

Downy Woodpecker (*Picoides pubescens*)

Hairy Woodpecker (*Picoides villosus*)

American Three-toed Woodpecker (*Picoides dorsalis*)

Black-backed Woodpecker (*Picoides arcticus*)

Northern Flicker (*Colaptes auratus*)

Pileated Woodpecker (*Dryocopus pileatus*)

Olive-sided Flycatcher (*Contopus cooperi*)

Western Wood-Pewee (*Contopus sordidulus*)

Eastern Wood-Pewee (*Contopus virens*)

Yellow-bellied Flycatcher (*Empidonax flaviventris*)

Alder Flycatcher (*Empidonax alnorum*)

Least Flycatcher (*Empidonax minimus*)

Hammond's Flycatcher (*Empidonax hammondii*)

Dusky Flycatcher (*Empidonax oberholseri*)

Pacific-slope Flycatcher (*Empidonax difficilis*)

Eastern Phoebe (*Sayornis phoebe*)

Say's Phoebe (*Sayornis saya*)

Great Crested Flycatcher (*Myiarchus crinitus*)

Eastern Kingbird (*Tyrannus tyrannus*)

Loggerhead Shrike (*Lanius ludovicianus*)

Northern Shrike (*Lanius excubitor*)

Yellow-throated Vireo (*Vireo flavifrons*)

Cassin's Vireo (*Vireo cassinii*)

Blue-headed Vireo (*Vireo solitarius*)

Warbling Vireo (*Vireo gilvus*)

Philadelphia Vireo (*Vireo philadelphicus*)

Red-eyed Vireo (*Vireo olivaceus*)

Gray Jay (*Perisoreus canadensis*)

Blue Jay (*Cyanocitta cristata*)

Clark's Nutcracker (*Nucifraga columbiana*)

Black-billed Magpie (*Pica hudsonia*)

American Crow (*Corvus brachyrhynchos*)

Common Raven (*Corvus corax*)

Horned Lark (*Eremophila alpestris*)

Purple Martin (*Progne subis*)

Tree Swallow (*Tachycineta bicolor*)

Violet-green Swallow (*Tachycineta thalassina*)

Northern Rough-winged Swallow (*Stelgidopteryx serripennis*)

Bank Swallow (*Riparia riparia*)

Cliff Swallow (*Petrochelidon pyrrhonota*)

Barn Swallow (*Hirundo rustica*)

Black-capped Chickadee (*Poecile atricapillus*)

Mountain Chickadee (*Poecile gambeli*)

Chestnut-backed Chickadee (*Poecile rufescens*)

Boreal Chickadee (*Poecile hudsonicus*)

Gray-headed Chickadee (*Poecile cinctus*)

Red-breasted Nuthatch (*Sitta canadensis*)

White-breasted Nuthatch (*Sitta carolinensis*)

Brown Creeper (*Certhia americana*)

House Wren (*Troglodytes aedon*)

Winter Wren (*Troglodytes hiemalis*)

Sedge Wren (*Cistothorus platensis*)

Marsh Wren (*Cistothorus palustris*)

American Dipper (*Cinclus mexicanus*)

Golden-crowned Kinglet (*Regulus satrapa*)

Ruby-crowned Kinglet (*Regulus calendula*)

Arctic Warbler (*Phylloscopus borealis*)

Bluethroat (*Luscinia svecica*)

Northern Wheatear (*Oenanthe oenanthe*)

Eastern Bluebird (*Sialia sialis*)

Mountain Bluebird (*Sialia currucoides*)

Townsend's Solitaire (*Myadestes townsendi*)

Veery (*Catharus fuscescens*)

Gray-cheeked Thrush (*Catharus minimus*)

Bicknell's Thrush (*Catharus bicknelli*)

Swainson's Thrush (*Catharus ustulatus*)

Hermit Thrush (*Catharus guttatus*)

Wood Thrush (*Hylocichla mustelina*)

American Robin (*Turdus migratorius*)

Varied Thrush (*Ixoreus naevius*)

Gray Catbird (*Dumetella carolinensis*)

Brown Thrasher (*Toxostoma rufum*)

European Starling (*Sturnus vulgaris*)

Eastern Yellow Wagtail (*Motacilla tschutschensis*)

Red-throated Pipit (*Anthus cervinus*)

American Pipit (*Anthus rubescens*)

Sprague's Pipit (*Anthus spragueii*)

Bohemian Waxwing (*Bombycilla garrulus*)

Cedar Waxwing (*Bombycilla cedrorum*)

Tennessee Warbler (*Oreothlypis peregrina*)

Orange-crowned Warbler (*Oreothlypis celata*)

Nashville Warbler (*Oreothlypis ruficapilla*)

Northern Parula (*Parula americana*)

Yellow Warbler (*Dendroica petechia*)

Chestnut-sided Warbler (*Dendroica pensylvanica*)

Magnolia Warbler (*Dendroica magnolia*)

Cape May Warbler (*Dendroica tigrina*)

Black-throated Blue Warbler (*Dendroica caerulescens*)

Yellow-rumped Warbler (*Dendroica coronata*)

Black-throated Green Warbler (*Dendroica virens*)

Townsend's Warbler (*Dendroica townsendi*)

Blackburnian Warbler (*Dendroica fusca*)

Pine Warbler (*Dendroica pinus*)

Palm Warbler (*Dendroica palmarum*)

Bay-breasted Warbler (*Dendroica castanea*)

Blackpoll Warbler (*Dendroica striata*)

Black-and-white Warbler (*Mniotilta varia*)

American Redstart (*Setophaga ruticilla*)

Ovenbird (*Seiurus aurocapilla*)

Northern Waterthrush (*Parkesia noveboracensis*)

Connecticut Warbler (*Oporornis agilis*)

Mourning Warbler (*Oporornis philadelphia*)

MacGillivray's Warbler (*Oporornis tolmiei*)

Common Yellowthroat (*Geothlypis trichas*)

Wilson's Warbler (*Wilsonia pusilla*)

Canada Warbler (*Wilsonia canadensis*)

Scarlet Tanager (*Piranga olivacea*)

Western Tanager (*Piranga ludoviciana*)

Spotted Towhee (*Pipilo maculatus*)

Eastern Towhee (*Pipilo erythrophthalmus*)

American Tree Sparrow (*Spizella arborea*)

Chipping Sparrow (*Spizella passerina*)

Clay-colored Sparrow (*Spizella pallida*)

Brewer's Sparrow (*Spizella breweri*)

Vesper Sparrow (*Pooecetes gramineus*)

Savannah Sparrow (*Passerculus sandwichensis*)

Baird's Sparrow (*Ammodramus bairdii*)

Le Conte's Sparrow (*Ammodramus leconteii*)

Nelson's Sharp-tailed Sparrow (*Ammodramus nelsoni*)

Fox Sparrow (*Passerella iliaca*)

Song Sparrow (*Melospiza melodia*)

Lincoln's Sparrow (*Melospiza lincolnii*)

Swamp Sparrow (*Melospiza georgiana*)

White-throated Sparrow (*Zonotrichia albicollis*)

Harris's Sparrow (*Zonotrichia querula*)

White-crowned Sparrow (*Zonotrichia leucophrys*)

Golden-crowned Sparrow (*Zonotrichia atricapilla*)

Dark-eyed Junco (*Junco hyemalis*)

Lapland Longspur (*Calcarius lapponicus*)

Smith's Longspur (*Calcarius pictus*)

Snow Bunting (*Plectrophenax nivalis*)

Rose-breasted Grosbeak (*Pheucticus ludovicianus*)

Indigo Bunting (*Passerina cyanea*)

Bobolink (*Dolichonyx oryzivorus*)

Red-winged Blackbird (*Agelaius phoeniceus*)

Eastern Meadowlark (*Sturnella magna*)

Western Meadowlark (*Sturnella neglecta*)

Yellow-headed Blackbird (*Xanthocephalus xanthocephalus*)

Rusty Blackbird (*Euphagus carolinus*)

Brewer's Blackbird (*Euphagus cyanocephalus*)

Common Grackle (*Quiscalus quiscula*)

Brown-headed Cowbird (*Molothrus ater*)

Baltimore Oriole (*Icterus galbula*)

Gray-crowned Rosy-Finch (*Leucosticte tephrocotis*)

Pine Grosbeak (*Pinicola enucleator*)

Purple Finch (*Carpodacus purpureus*)

Red Crossbill (*Loxia curvirostra*)

White-winged Crossbill (*Loxia leucoptera*)

Common Redpoll (*Acanthis flammea*)

Hoary Redpoll (*Acanthis hornemanni*)

Pine Siskin (*Spinus pinus*)

American Goldfinch (*Spinus tristis*)

Evening Grosbeak (*Coccothraustes vespertinus*)

House Sparrow (*Passer domesticus*)

CHAPTER THREE

Waterfowl Conservation Planning

SCIENCE NEEDS AND APPROACHES

Stuart M. Slattery, Julienne L. Morissette, Glenn G. Mack, and Eric W. Butterworth

Abstract. The western boreal forest (WBF) is the second most important duck breeding area in North America. Once thought to be relatively pristine, WBF habitat is undergoing rapid change due to industrial activity (e.g., commercial forestry, oil and gas exploration/extraction, agricultural expansion) and climate change. Our understanding of waterfowl in the WBF is limited and thus the effects of these human-caused habitat alterations on waterfowl populations are largely unknown. A better understanding of the spatial scale, permanency, and intensity of habitat change at which carrying capacity becomes reduced will help differentiate between real and perceived threats, which is critical for further focusing conservation. Maintaining the ability of boreal landscapes to sustain waterfowl populations in perpetuity, however, will require interdisciplinary collaboration. Ultimately, conservation activities will take place within a complex ecological and socio-economic landscape, which will require a strong commitment from all stakeholders, including industry, First Nations, governments, academics, and NGOs, to achieving conservation of boreal landscapes that encompass wetlands and other waterfowl habitat. In this paper, we review the status and population trends of duck populations in the WBF, review threats to carrying capacity, and provide an initial conceptual framework for science-based conservation of WBF ducks.

Key Words: conservation planning, science priorities, waterfowl, western boreal forest, wetlands.

The North American Waterfowl Management Plan (NAWMP) recognizes the western boreal forest (WBF; Fig. 3.1) as the second most important region on the continent for breeding waterfowl (North American Waterfowl Management Plan, Plan Committee 2004). NAWMP was established in 1986 to guide waterfowl conservation in the United States, Canada, and Mexico through collaborations among government and non-governmental organizations. The plan's overall goal is to sustain North American waterfowl populations at 1970s levels with science-based approaches to conserving both wetland and upland habitats used by breeding, molting, staging, and wintering birds (North American Waterfowl Management Plan, Plan Committee 2004). In 2007, the NAWMP Assessment Committee recommended that boreal conservation actions be

Slattery, S. M., J. L. Morissette, G. G. Mack, and E. W. Butterworth. 2011. Waterfowl conservation planning: science needs and approaches. Pp. 23–40 *in* J. V. Wells (editor). Boreal birds of North America: a hemispheric view of their conservation links and significance. Studies in Avian Biology (no. 41), University of California Press, Berkeley, CA.

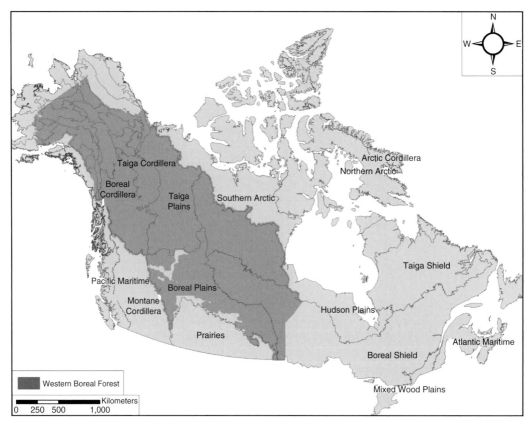

Figure 3.1. Western boreal forest boundary, including delineation of ecozones.

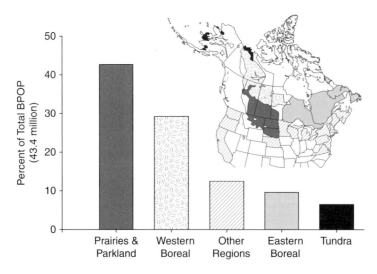

Figure 3.2. Long-term average distribution of total ducks counted on breeding grounds in North America. Data are from the Waterfowl Breeding Population and Habitat (BPOP) surveys (Traditional Survey Area, 1960–2007; Eastern Survey Area (Eastern Canada and Maine), 1990–2006) and state breeding season waterfowl survey data reported in USFWS (2007). States had different survey durations but generally overlapped with BPOP survey lengths (mean [sd] = 28 [14] years). Inset map indicates regions where data were collected, with colors corresponding to bar colors.

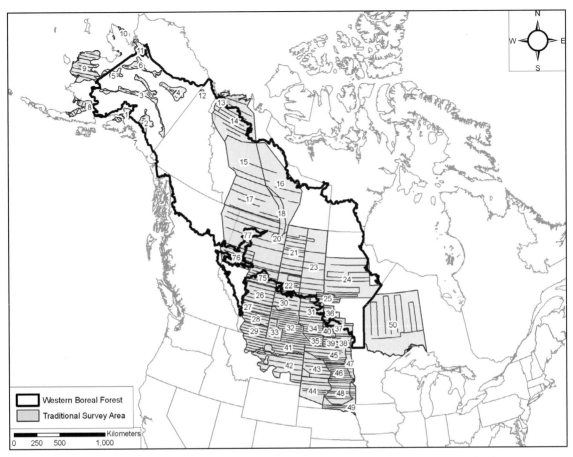

Figure 3.3. Traditional Survey Area strata boundaries and transects of the Waterfowl Breeding Population and Habitat Survey (U.S. Fish and Wildlife Service and the Canadian Wildlife Service, 1987). Numbers indicate strata identification.

more explicitly tied to NAWMP, indicating mainstream support for WBF conservation within the waterfowl research and conservation community.

The WBF is a high-priority area (North American Waterfowl Management Plan, Plan Committee 2007) because this region (Fig. 3.1), contains about 29% of the ducks counted in North America during the breeding season (Fig. 3.2) and annually supports 12–15 million ducks each spring, representing 23 species (Appendix 3.1). Annual estimates for WBF populations are derived from the Waterfowl Breeding Population and Habitat Surveys (hereafter BPOP surveys), conducted since 1955 in core western breeding areas (Fig. 3.3; hereafter referred to as the Traditional Survey Area or TSA) (U.S. Fish and Wildlife Service and Canadian Wildlife Service 1987); waterfowl in the WBF comprise up to 41% of the annual TSA population estimate (Appendix 3.1). Eight of the core species counted within the TSA have >50% of their TSA breeding

season populations located in the WBF (Table 3.1), and about 85% of WBF ducks occur in Canada, where the top four numerically most important species are below NAMWP goals (Table 3.1). Reasons for this status are largely unknown, but limited evidence suggests that important factors may be acting on boreal breeding grounds, at least for scaup (Afton and Anderson 2001). However, these declines may be offset by increases in other species and regions because total duck populations in the WBF generally have been relatively stable, particularly compared to the prairie biome (Fig. 3.4).

Large numbers of waterfowl use the WBF, partly because it is within a biome containing the greatest number of wetlands (forested and non-forested) and lakes in the world (Foote and Krogman 2006). Wetlands occupy between 25 and 50% of the landscape (Vitt 1994). Until relatively recently, conserving boreal forest waterfowl habitat was considered a lower priority than conserving waterfowl habitat

TABLE 3.1

North American Waterfowl Management Plan (NAWMP) population goals for ducks in the western boreal forest (WBF) and population status relative to those goals

Species[a]	% in WBF[b]	NAWMP Goals[c]	US Status[d]	CN Status[d]
MALL	28.7	1,965,753	61	−25
CANV	34.3	143,091	8	76
AMWI	53.3	1,268,403	53	−32
AGWT	55.4	730,182	155	38
SCAU	67.2	3,814,483	−14	−58
BUFF	67.8	522,425	−33	17
GOLD	76.7	297,882	5	88
SCOT	79.9	950,328	−19	−50
RNDU	83.0	342,144	34	135
MERG	90.1	261,318	188	149
OVERALL[e]	29.1	12,255,067	31	−17

[a] Species: AGWT = American Green-winged Teal *Anas crecca*; AMWI = American Wigeon *Anas americana*; BUFF = Bufflehead *Bucephala albeola*; CANV = Canvasback *Aythya valisineria*; GOLD = Common Goldeneye *Bucephala clangula* and Barrow's Goldeneye *Bucephala islandica*; MALL = Mallard *Anas platyrhynchos*; MERG = Common Merganser *Mergus merganser*, Hooded Merganser *Lophodytes cucullatus*, and Red-breasted Merganser *Mergus serrator*; RNDU = Ring-necked Duck *Aythya collaris*; SCAU = Lesser Scaup *Aythya affinis* and Greater Scaup *Aythya marila*; SCOT = Black Scoter *Melanitta nigra*, Surf Scoter *Melanitta perspicillata*, and White-winged Scoter *Melanitta fusca*.

[b] Only species with greater than 25% of their Traditional Survey Area (TSA) population in the WBF were included.

[c] NAWMP population goals are typically 1970s average populations (North American Waterfowl Management Plan Committee 2004). However, only 1977–1979 were used for the U.S. because of a change in aircraft in Alaska in 1977 resulted in increased visibility thereafter (Hodges et al. 1996), which are not corrected for in the data.

[d] Canada and United States were separated for status analyses because of differences in population trends among regions (Zimpfer et al. 2009). Status is the percent difference between NAWMP population goals and mean population estimates from 2000–2009 within each region.

[e] Includes all duck species.

elsewhere because its productivity was not fully recognized (Lynch 1984). In addition, wetlands in the WBF were not perceived as sufficiently threatened by development activities or other causes of loss (North American Waterfowl Management Plan, Plan Committee 2004), particularly compared to the prairies, which in some areas have seen up to 70% of wetlands lost (http://atlas.nrcan.gc.ca/site/english/learningresources/theme_modules/wetlands/index.html). However, boreal aquatic systems are facing a mounting suite of threats (Schindler 1998, Foote and Krogman 2006) due to increasing direct and indirect anthropogenic activities (Schneider 2000, Hobson et al. 2002).

While ecological research on ducks and wetlands in boreal regions is increasing (e.g., Brook et al. 2005, Walker and Lindberg 2005, Hornung and Foote 2006, Paszkowski and Tonn 2006, Schmidt et al. 2006, Walsh et al. 2006, Corcoran et al. 2007), conservation planning is challenged because the effects of increasing human disturbance on

the ecology of wetlands and waterfowl (e.g., vital rates) are largely not quantified (but see Pierre et al. 2001; Gurney et al. 2005 for examples). If we are to progress, a better understanding of which disturbances threaten waterfowl habitat and, more specifically, the spatial scale, permanency, and intensity of changes that reduce WBF carrying capacity is essential to advance conservation programs.

While conservation action in the WBF will be critical to achieving goals of NAWMP and other conservation initiatives, important uncertainties limit that action. Those uncertainties generally are captured by four key questions:

1. Where are the birds?

2. Why are they there?

3. How might waterfowl populations change in response to anthropogenic activity?

4. What can be done to minimize that change?

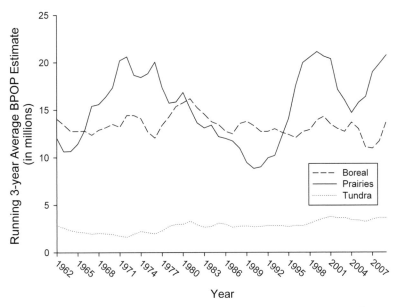

Figure 3.4. Breeding season duck population trends for three biomes in North America. The Boreal biome (dashed line) includes the Alaska boreal region, Boreal Plains, Taiga Plains, Boreal Shield, Taiga Shield, Taiga Cordillera, and the Boreal Transition ecozones. The Prairies biome (solid line) includes the Prairie potholes, Parklands, Northern Rockies, Badlands and High Prairies. The Tundra (dotted line) includes the Alaska tundra and southern arctic ecozones.

The objective of this chapter is to foster greater dialogue within the conservation community about conserving both terrestrial and aquatic ecosystems within the WBF by exploring how the questions above influence waterfowl conservation. Up to now, this dialogue has largely underemphasized the need to conserve aquatic ecosystems.

WHERE ARE THE BIRDS?

Understanding spatial and temporal variation in distribution and abundance of a species is an essential first step in targeting habitat conservation efforts (Margules and Pressey 2000) aimed at continental, regional, and local scales. Waterfowl conservation is no exception (Fast and Berkes 1999). This knowledge helps prioritize habitats and increases the efficacy of conservation action by permitting design of conservation approaches that are appropriate for waterfowl distribution patterns. For example, many efforts at preserving biodiversity or individual species have tended to focus on "areas of key importance" for conservation, unique areas that have a high degree of biodiversity or high abundance of species of interest (Margules and Pressey 2000). Other efforts focus on protecting areas representing key ecological features that occur across the larger landscape.

However, the primary aim of NAWMP is to ensure that the landscape is capable of sustaining historic levels of waterfowl populations. Therefore, to be successful, conservation actions must positively influence high proportions of regional or continental populations. Our preliminary analyses of BPOP survey data indicate that even when scaup populations were high in the 1970s, >75% of their western boreal population occurred at densities of <10 pairs per square mile, and 50% were in regions of <5 pairs per square mile. Spatial distribution patterns of other waterfowl taxa have not yet been fully examined, but it seems that while the western boreal forest accounts for a large portion of the continental breeding population of some species, they tend to occur at low densities across this large area. Based on such distribution patterns, the paradigm of protecting high-density areas, while a fundamental component of a conservation approach, probably impacts too few birds to sustain WBF waterfowl populations at target levels. If so, programs with broad spatial impact will be critical to effective habitat conservation in the WBF. We argue that the deployment of human and financial resources toward developing and implementing conservation must be guided by an understanding of how birds are distributed across the landscape of interest. Where uncertainties remain, assumptions must be clearly articulated.

Answering "Where are the birds" is also tightly linked to the question "Where are the wetlands?" These include bogs, fens, swamps, marshes, and open water basins. Without knowledge of the location of wetlands, it is difficult to assess the potential impact of landscape changes at a broad spatial scale and implement efforts to conserve them. However, until recently there were few attempts to accurately map wetland distribution in the boreal forest using an ecologically based classification system such as the Canadian Wetland Classification System (National Wetlands Working Group 1997). As large-scale habitat data sets including wetland maps become more available, potentially powerful tools such as predictive habitat distribution models (Guisan and Zimmermann 2000) can be developed to assist conservation planning for waterfowl. For example, despite some shortcomings (see Afton and Anderson 2001), the BPOP survey is currently one of the best broad-scale and long-term data sets available for any wildlife population. Using these survey data with appropriately scaled habitat data may allow identification of key associations between waterfowl densities and landscape features. These associations can then be used to produce maps of predicted distribution and abundance of waterfowl species across the WBF, potentially identifying both long-term average distribution patterns and temporal trends in spatial distributions. When combined with maps of current and future habitat threats, these models would help managers focus conservation planning efforts on areas that are both important to ducks and at sufficient risk of loss to merit conservation investment.

WHY ARE THEY THERE?

Understanding the distribution pattern of waterfowl is a critical first step in conservation because such analyses may guide us to key waterfowl landscapes or conservation philosophies appropriately scaled to densities. In some cases, this knowledge may be sufficient to proceed with conservation action, particularly when intact landscapes are threatened or strategic conservation opportunities are pressing. But an understanding of waterfowl distribution may do little to identify the types of conservation programs that are likely to succeed in those landscapes where sufficiently different

competing hypotheses about limiting factors exist. In such cases, answering "Why are they there?" may lead to more effective conservation, and require an understanding of both biotic and abiotic factors influencing settling patterns (e.g., wetland density, wetland type, food availability) and variation in vital rates across habitats during pre-breeding, nesting, brood-rearing, and staging periods. As with other organisms, the mechanisms driving distribution patterns of waterfowl populations may also be highly variable, both spatially and temporally, so capturing the appropriate conservation action at the appropriate spatial scale will be a challenge (Elkie and Rempel 2001).

A rich history of research on waterfowl settling patterns and vital rates exists from the Prairie Pothole Region. However, applicability of those results to understanding boreal waterfowl ecology is unknown but is likely limited because the ecology of a single species can vary from one region to the next even within biomes (Guyn and Clark 2000, Lake et al. 2006). For example, survival rates of boreal waterfowl populations could differ from prairie populations due to increased energetic demands of migration (Alisauskas and Ankney 1992) and potentially greater vulnerability to harvest (U.S. Fish and Wildlife Service 2003). In addition, it has been hypothesized that mortality of female breeding ducks is lower in boreal regions because of lower human influence on the landscape and lower predation rates (Sargeant and Raveling 1992) associated with a shorter nesting season. Data from prairie studies have allowed waterfowl managers to develop models predicting the productivity and abundance of waterfowl using parameters such as wetland permanence, size, upland vegetation, surrounding land use, patch size, vegetation structure, and wetland class (e.g., Krapu et al. 1983, Anderson and Titman 1992, Greenwood et al. 1995, Guyn and Clark 2000, Devries et al. 2004). However, ecological conditions at northern latitudes are likely different from those in southern regions and species may use different life history strategies to cope with shorter breeding seasons and different food sources when breeding in the north (Paszkowski and Tonn 2000, Lake et al. 2006, Schmidt et al. 2006).

Answering the question "Why are they there?" can also lead to a better understanding of factors influencing wetland productivity, allowing us to establish estimates for the natural range of variation in productivity against which to judge effects of anthropogenic landscape changes important to ducks and refine working hypotheses about how those changes might limit waterfowl populations. Wetland productivity is influenced by hydrological and biotic processes affecting the flow of water and nutrients through the landscape (National Wetlands Working Group 1997), and we are just beginning to understand that the hydrologic connectivity among wetlands is high and more complex than originally thought (Devito et al. 2005). This complexity implies that effects of industrial activity could be more difficult to predict for WBF wetlands than for wetlands in other biomes. However, information about hydrologic connectivity across the boreal forest is quite limited (Foote and Krogman 2006, but see Devito et al. 2005, Petrone et al. 2007). Improving our knowledge of factors affecting wetland productivity, particularly when examined in the context of land-use changes known to reduce duck populations, will likely lead to more effective habitat conservation.

When planning for WBF conservation, we also need to recognize that contemporary spatial distributions and abundances of waterfowl in the boreal forest provide limited information regarding population dynamics with which to guide biological planning efforts. Unfortunately, we have little demographic information from the boreal forest on which to base predictions about how bird populations are regulated and how they might respond to habitat change (Niemi et al. 1998). However, examining spatial variation in population trends and community persistence over broad spatial scales may indicate where changes in carrying capacity have occurred (Bethke 1993). Indeed, identification of regions where populations have increased, remained stable, or declined would permit comparative studies on processes underlying the respective local population trajectories, identifying potential targets for management actions and facilitating research on effects of anthropogenic disturbances.

HOW MIGHT WATERFOWL POPULATIONS CHANGE IN RESPONSE TO ANTHROPOGENIC ACTIVITY?

Human activity is changing WBF waterfowl habitat directly through a variety of industrial and recreational activities and indirectly through climate change. However, the impact of these disturbances on waterfowl populations at various spatial scales is unclear. Natural disturbances such as forest fires, insect outbreaks, wind, and ice storms are an intrinsic part of boreal systems that contribute to and help maintain their heterogeneity (Niemi et al. 1998). Thus boreal ecosystems and species are expected to be well adapted to natural disturbances (Niemi et al. 1998, Schmiegelow and Mönkkönen 2002, Lemelin et al. 2007) and to have a certain resilience to change. However, we do not fully know how anthropogenic activity compares to natural disturbance in terms of intensity and spatial extent of perturbation, and more particularly if aquatic ecosystems are more or less resilient than terrestrial ecosystems where the bulk of research on emulating natural disturbance has occurred (e.g., Hobson and Schieck 1999, Lindenmayer and Noss 2006). Therefore, a top priority for boreal waterfowl science is identifying which landscape changes threaten WBF carrying capacity, and more specifically at what spatial scale, permanency, and intensity they become limiting to ducks.

The growing occurrence of anthropogenic habitat change in the WBF due to timber harvesting, oil and gas exploration and extraction, mining, hydroelectric development, and recreation, particularly when combined with effects of climate change, could jeopardize waterfowl carrying capacity. These impacts could be exacerbated in permafrost regions, where wetland persistence requires frozen basin bottoms (Riordan et al. 2001). The individual and combined effects of these disturbances are not well studied, particularly for wetlands and waterfowl. Here we provide a brief overview of some key anthropogenic disturbances and potential impacts on western boreal waterfowl communities.

Commercial Forestry

By area, logging is currently the primary direct human-induced change in boreal forests (Niemi et al. 1998), although in some areas, tree removal by commercial forestry is exceeded by tree removal for petroleum extraction activities (Schneider 2000). The boreal shield and boreal plain ecoregions (Fig. 3.1) contain the majority of the commercial forest in the WBF. Wetlands with open water are generally protected from the direct effects of forest activities by riparian management guidelines (e.g., buffers) and rules regarding in-block road placement, application of pesticides, and other issues. However, the ecological basis for and efficacy of these guidelines has been the subject of debate.

Forest harvesting activities could potentially affect how and when birds use wetlands, where they nest, or the behavior of predators. However, we are aware of few studies that have examined the effects of forest harvesting on waterfowl, particularly for ground- or overwater-nesting species (but see Pierre et al. 2001). It has been hypothesized that short return intervals for harvesting could lead to an overall homogenization of and decrease in stand ages, plus subsequent loss of snags and potential cavity trees (Angelstam 1992) used by species such as Bufflehead (*Bucephala albeola*) and goldeneye (*Bucephala* spp).

Petroleum Extraction

Road-building and seismic exploration for oil and gas extraction create linear disturbances resulting in both habitat loss and fragmentation. Since much exploration is carried out in winter, seismic lines often run through wetlands. In Alberta, Canada, there is also currently no limit to the cumulative density of seismic lines allowed (Schneider et al. 2003), and in 1999 100,000 km (63,000 mi) of seismic lines had been approved (Lee and Boutin 2006). Densities of seismic lines ranging from 1.5 to 10 km per km^2 were also observed in 98% of the townships in northeast Alberta by 2001. In summer, water is needed to run pumpjacks; this water is taken from wetlands and freshwater aquifers. These pumpjacks and compressor stations are often placed near wetlands, which can also result in habitat loss. The effects of these disturbances on waterfowl, however, have not yet been quantified at either community or population level.

Added to typical oil and gas operations, oil sands extraction accounts for approximately 15% of Canada's oil production. In Alberta, the oils sands region underlies 140,200 km^2 (54,132 mi^2). As of June 2009, mineral rights agreements have been issued for 82,542 km^2 (31,870 mi^2; http://www.energy.alberta.ca/OilSands/791.asp), an area about the size of Maine. As of 2009, surface mines and associated footprints had disturbed 686 km^2 (262 mi^2) of land. However, half of this area had existed as wetlands including peatlands and open

water wetlands (Carlson et al. 2009), and similar proportions of wetlands account for the remaining oil sands region. Wetlands are lost in the extraction process, while artificial habitats in the form of tailings ponds also are created (Bishay and Nix 1996). Some of these intentionally created wetlands of consolidated tailings effluent are capable of supporting low-diversity benthic communities (King and Bendell-Young 2000). However, they appear to be poor-quality habitats, resulting in reduced growth of ducklings relative to reference ponds (Gurney et al. 2005), and so likely do not replace lost wetlands. Ducklings also ingest sediment and grit, which may increase exposure to oil sands contaminants (Gurney et al. 2005), and negatively affect growth (Szaro et al. 1981). Poor growth may lead to small body size, decreased survival, and poor recruitment (Rhymer 1988, Cox et al. 1998, Christensen 1999, Hill et al. 2003). While these studies were not carried out at the population level, results do suggest that tailings ponds may be ecological traps for waterfowl.

Agriculture

The southern fringe of the western boreal forest has experienced a high rate of conversion to agriculture. For example, from 1966 to 1994, Saskatchewan's boreal plain was converted to agriculture at 0.89% per year, three times the global deforestation rate (Hobson et al. 2002). The impact of this conversion on wetland ecosystems and boreal waterfowl populations is unknown, but in prairie or parkland landscapes, agricultural expansion often leads to wetland drainage and permanent conversion of forested habitat. In some cases, economic incentives promote the conversion of forests to agriculture (Hobson et al. 2002). Relative to forestry, conversion to non-forest cover results in static landscape configurations (Schmiegelow and Mönkkönen 2002). The full range of ecological implications of agricultural conversion has not been well examined in boreal systems, but examples include: higher predator densities (e.g., corvids, skunks), potentially resulting in higher nest predation rates (Andren 1992, Bayne and Hobson 1997); altered water chemistry due to increased nutrient inputs (Houlahan and Finlay 2004); and subsequent changes in food web dynamics. To our knowledge, no published studies have specifically examined the effects of conversion of boreal forests to agriculture on wetland-associated birds.

Roads

Roads are known to alter movement patterns of wildlife, increase levels of animal mortality due to collision with vehicles, result in both habitat loss and fragmentation, increase human access to the backcountry, and increase sedimentation at road crossings, which can impair wetland connectivity and productivity (Trombulak and Frissell 2000). Trombulak and Frissell (2000) and Forman and Alexander (1998) provide a detailed review of the ecological effects of roads; however, there are few studies that specifically examine road effects on boreal waterfowl or wetlands. A study of the effects of roads on species richness of mammals, herptiles, and birds at Ontario wetlands found a negative correlation between density of paved roads and species richness (Findlay and Houlahan 1997). However, another study showed that distance to nearest road had no effect on the presence of brood rearing in Lesser Scaup (Fast et al. 2004). In the same study, scaup pairs and nests were frequently observed at borrow pits, created when building roads, but broods were found almost exclusively on natural wetlands (Fast and Berkes 1999). In southern Ontario, time since disturbance has been shown to play a significant role in the productivity of wetlands (Findlay and Bourdages 2000). In addition, lag effects were observed with respect to effects of roads on birds, herptiles, and plants, with some species losses evident after eight years but others not detectable until after several decades (Findlay and Bourdages 2000). Since access for industrial operations represent much of the road expansion in the WBF and are relatively recent, long-term monitoring along with comparative studies will be useful for establishing whether roads lead to detrimental effects for waterfowl.

Cumulative Effects

Finally, combined and potentially cumulative effects of anthropogenic habitat disturbance on waterfowl are unknown. These effects could be through direct habitat loss due to hydrologic change and terrestrial disturbances, or through subtle indirect effects such as changes in predator behavior. Research on cumulative effects in Alberta has documented the expansion of coyote and other generalist predators into the boreal region with increasing human disturbances (Nielsen et al.

2007), and greater use of linear features by wolves (James and Stuart-Smith 2000); thus, human disturbances may also increase accessibility by predators to adults, eggs, and ducklings. This top-down hypothesis remains untested and, to date, there are no published estimates of breeding season mortality in these areas.

Climate change is an effect of human activity that occurs both inside and outside the WBF that may complicate understanding and management of locally occurring human activity. Warming due to climate change is expected to be greatest in boreal regions (Pastor et al. 1998), likely also changing natural disturbance regimes (Drever et al. 2006), precipitation:evaporation ratios (Schindler 1998), and potentially exacerbating the effects of other human disturbances. In the boreal plain, potential evapotranspiration exceeds precipitation in most years, with wet years occurring on a 10–15 year cycle (Price et al. 2005). Since evapotranspiration dominates the hydrologic cycle in the WBF, the region is vulnerable to wetland loss due to climate change (Schindler 1998, Devito et al. 2005). In addition, where there is permafrost, warming trends could result in melting of permafrost, potentially altering hydrologic processes (Camill and Clark 1998) and food webs important for ducks (Corcoran et al. 2007). For example, evidence from Alaska (Riordan et al. 2006) and Siberia (Smith et al. 2005) indicated long-term declines in wetland surface area apparently linked to climate-induced degradation of permafrost. The spatial extent of this drying trend, particularly in the Canadian boreal forest, is unknown, but the implications are clear—areas subject to the greatest increase in temperatures will likely have the greatest wetland loss, and hence the most impact on waterfowl populations. Though we cannot readily predict how climate change will interact with other human disturbances (Chapin et al. 2004), developing a better understanding of such relationships will be key to managing impacts.

Given these many sources of habitat change, limited empirical data to support specific effects on waterfowl, and the accepted need for habitat conservation despite uncertainties, we offer two main working hypotheses about how habitat changes influence waterfowl populations (Ducks Unlimited Canada 2010). First, we hypothesize that open water wetlands are the most limiting habitat feature, and that these are subject to reduced abundance and, perhaps more important, quality by (a) damming and direct alteration by roads and other infrastructure, (b) sedimentation from roads and other activities that remove forest cover, (c) altered nutrient and water yield off sites with forest cover removed, (d) thawing of porous basin bottoms and associated changes in food webs due to climate change, and (e) thawing of permafrost induced by industrial exploration and infrastructure. We propose that these changes could result in bottom-up effects on all guilds of waterfowl through a reduction in per capita food availability, leading to fewer birds settling, fewer birds breeding, and lower survival of hens and their ducklings.

The second working hypothesis assumes that availability of upland nesting habitat is not limiting, but anthropogenic changes to uplands may reduce upland quality through increased habitat fragmentation. We hypothesize that (a) creation of travel corridors from roads, seismic lines, and other forest openings and (b) changes in vegetation structure–associated forest cover removal may result in increased predation rates due to immigration of new predator species, improved hunting success of native boreal predators, or growth of predator populations due to more abundant alternate prey. This top-down limitation is hypothesized to mainly affect ground-nesting waterfowl, and demographic implications are reduced hen, nest, and duckling survival that ultimately reduces local duck populations and/or creates population sinks. These negative demographic effects may be temporary if vegetation regrows, and may also have a positive effect at some stages of regeneration.

These working hypotheses are largely untested and are not necessarily mutually exclusive. However, we suggest that they can form the conceptual framework guiding future research on the relationships between anthropogenic landscape change and waterfowl populations.

WHAT CAN BE DONE TO MINIMIZE THAT CHANGE?

Understanding relationships between anthropogenic activities and waterfowl distribution, abundance, and demography is important to developing strategies for ensuring waterfowl populations are sustained at NAWMP goal levels. Because we know little about how anthropogenic activity in the western boreal forest impacts waterfowl, the

working hypotheses above are a first attempt at articulating potential relationships in a conceptual framework that is not only useful, but is perhaps imperative for focusing conservation design. However, these hypotheses are based on limited empirical data and expert opinion. This situation creates a climate of uncertainty regarding conservation strategies in the face of a growing suite of challenges. Despite uncertainties, scientists have begun to identify and promote strategies for maintaining the ecological resilience of boreal systems (Schmiegelow and Mönkkönen 2002, Chapin et al. 2004, Foote and Krogman 2006). Here, ecological resilience is defined as the capacity of an ecosystem to absorb disturbance and undergo change while maintaining its essential functions, structures, identity, and feedbacks (Walker et al. 2005, Drever et al. 2006).

Maintaining ecological resilience requires clearly defining goals for ecosystem function at various spatial and temporal scales, particularly identifying parameters of interest, their range of acceptable values, and patterns of variation, that is, setting ecological benchmarks (Holling 1973, Schmiegelow and Mönkkönen 2002, Drever et al. 2006). These benchmarks can then be used to judge performance of conservation and management actions. Setting ecological benchmarks, however, requires reference areas where ecological processes ideally are free from human impact. Given the limited information about basic waterfowl ecology in the boreal forest, let alone the interaction of human activities and waterfowl-related ecological processes, it is clear that measuring ecological resilience, together with achieving and maintaining it, has many mechanistic knowledge gaps. However, suggested activities for achieving conservation and maintaining ecological resilience of boreal systems have included: (1) establishing large ecological reserves (Lindenmayer et al. 2006); (2) maintaining the balance of ecological functions on the rest of the landscape (Schmiegelow and Mönkkönen 2002, Lindenmayer et al. 2006) and restoring areas where these have been lost (Foote and Krogman 2006); and (3) establishing and implementing policies that support these approaches (e.g., Foote and Krogman 2006, Lindenmayer and Noss 2006) within an adaptive management framework (see Lee 1993, Lancia et al. 1996). Given the apparent low density of waterfowl in the WBF on average, we suggest that if restoration is deemed necessary, it should be confined to places that previously held a high density of waterfowl. We suggest that the majority of conservation effort should be focused on retaining current ecosystem function, which might yield greater long-term return from limited conservation resources.

In parts of the boreal forest already allocated for development, substantial ecological reserves are unlikely to be established. It follows that outside of reserves, conservation activities on the working landscape will become critical. Cooperation is required between conservation planners and industry to develop and test management approaches that both maintain waterfowl carrying capacity and fit into industrial operations at various planning scales. Current approaches to landscape-level conservation planning include watershed-based planning, minimizing road networks, maintaining hydrologic function (Schindler 1998), and planning for spatial and temporal patterns of forest harvesting that emulate natural patterns (Hunter 1993, Niemela 1999, Drever et al. 2006, Lindenmayer et al. 2006). To date these approaches do not specifically incorporate strategies for waterfowl and must evolve as research continues to build on the biological foundation provided by working hypotheses, reducing basic uncertainties regarding the responses of waterfowl across a range of ecological or hydrological contexts.

Conservation approaches in working landscape often use the term "minimizing impacts" to describe intended outcomes, particularly with government and industry. However, this term is open-ended, with limited ability to establish performance criteria. While any reform that slows the rate of detrimental habitat change is a type of conservation gain, a critical associated question is "How much reduction of impact is enough to achieve conservation goals?" To result in sufficient waterfowl habitat function to meet NAWMP goals, impacts need to be sustained at a level that ensures carrying capacity, and hence demographic rates, remain high enough on average to sustain target population levels. Therefore, these population levels need to be used to define targets for objective-based beneficial management practices (National Ecological Assessment Team 2006).

While developing biologically based targets of conservation is a difficult first step, implementing conservation programs faces many challenges, including both a lack of policies that provide avenues for protection of ecological resources and the existence of many policies that promote development. Often ecological values are managed in competition with industrial use rather than in collaboration. Also, different industries continue to be regulated under separate policies (Schneider et al. 2003). This disconnection creates a situation that facilitates increased cumulative effects that could be minimized by joint planning. For example, reducing road networks through joint planning of mineral and lumber extraction activities has been suggested, but there are many government and industry hurdles to this process (Schneider 2002).

Clearly, the list of information needs and policy implications for conserving boreal habitat is long, and changes to this landscape are occurring more rapidly than information gaps can be filled through directed research. Conservation activities should therefore proceed with an adaptive management approach. Adaptive management is the practice of learning while doing or developing management actions based on current understanding and assumptions, but treating the implementation of that action as an experiment designed to reduce key uncertainties (Holling 1978, Walters 1986, Walters and Holling 1990). This approach includes establishing clearly defined quantitative predictions based on working hypotheses that include measures of success. Ideally, these measures of success reflect maintenance of ecological resilience, and for waterfowl, they would be grounded in demographic rates as indicators of carrying capacity.

However, measuring demographic rates precisely is exceptionally difficult in this landscape. Therefore, one approach in the near term might be to determine environmental factors (e.g., measures of ecological processes, sets of ideal habitat conditions), measures of distribution, abundance, community structure, or indices waterfowl productivity that can serve as reliable correlates for demographic rates (*sensu* Margules and Pressey 2000). Even identifying these surrogates will be difficult because linkages between waterfowl demography and correlates, at least initially, will have to be based largely on limited knowledge while empirical data are being collected. In addition, strategies to monitor these demographic proxies and landscape changes thought to influence them are required across broad spatial scales. Such monitoring programs will permit us to track progress of conservation actions relative to carrying capacity goals.

Maintaining the ability of boreal landscapes to sustain waterfowl populations in perpetuity will require interdisciplinary collaboration to fill substantial information needs. These needs include (1) obtaining information about how boreal ducks respond to habitat change, and where appropriate underlying ecological processes, on which to build conservation programs; (2) identifying and prioritizing areas for targeting delivery of conservation action; and (3) establishing ecologically based performance objectives for best management practices. Ultimately, we must also evaluate and improve the efficacy of our actions within a complex ecological and socioeconomic landscape, which requires implementation of a long-term adaptive process. Finally, and most importantly, conservation of boreal waterfowl habitat can never be achieved without a strong commitment among stakeholders—industry, First Nations, governments, academics, and NGOs—to developing and realizing a common vision for future boreal landscapes that includes healthy wetland ecosystems.

ACKNOWLEDGMENTS

We would like to acknowledge the assistance of T. Lanson, M. Spearman, and M. Robin in the preparation of this manuscript. Reviews by Dave Howerter, Karla Guyn, Jeff Wells, SAB editors, and two anonymous reviewers greatly improved the quality of the manuscript. Mike Anderson, Jim Devries, Karla Guyn, Mark Gloutney, Bob Hayes, Shannon Hazard, Dave Howerter, Fritz Reid, Al Richard, Chris Smith, Sean Smythe, and Gary Stewart contributed tremendously to the conceptual development of this paper. We also thank the U.S. Fish and Wildlife Service, Environment Canada, and all other contributing agencies for allowing us access to the BPOP data. Funding for participation in the 2007 North American Ornithological Congress was provided by Ducks Unlimited Canada, Western Boreal Program. We thank Jeff Wells for the invitation to participate in the North American Ornithological Congress Boreal Bird Session.

LITERATURE CITED

Afton, A. D., and M. G. Anderon. 2001. Declining scaup populations: a retrospective analysis of long-term population and harvest survey data. Journal of Wildlife Management 65:781–796.

Alisauskas, R. T., and C. D. Ankney. 1992. The cost of egg laying and its relationship to nutrient reserves in waterfowl. Pp. 30–61 *in* B. D. J. Batt, A. D. Afton, M. G. Anderson, C. D. Ankney, D. H. Johnson, J. A. Kadlec, and G. L. Krapu (editors), Ecology and management of breeding waterfowl. University of Minnesota Press, Minneapolis, MN.

Anderson, M. G., and R. D. Titman. 1992. Spacing patterns. Pp. 251–289 *in* B. D. J. Batt, A. D. Afton, M. G. Anderson, C. D. Ankney, D. H. Johnson, J. A. Kadlec, and G. L. Krapu (editors), Ecology and management of breeding waterfowl. University of Minnesota Press, Minneapolis, MN.

Andren, H. 1992. Corvid density and nest predation in relation to forest fragmentation: a landscape perspective. Ecology 73:794–804.

Angelstam, P. 1992. Conservation of communities: the importance of edges, surroundings and landscape mosaic structure. Pp. 9–70 *in* L. Hansson (editor), Ecological principles of nature conservation. Elsevier, London, UK.

Bayne, E. M., and K. A. Hobson. 1997. Comparing the effects of landscape fragmentation by forestry and agriculture on predation of artificial nests. Conservation Biology 11:1418–1429.

Bethke, R. W. 1993. Geographical patterns of persistence in duck guilds. Oecologia 93:102–108.

Bishay, F. S., and P. G. Nix. 1996. Constructed wetlands for treatment of oil-sands wastewater. Technical Report 5. Prepared for Suncor Oil-Sands Group. EVS Environmental Consultants, North Vancouver, BC.

Brook, R. W., D. C. Duncan, J. E. Hines, S. Carriere, and R. G. Clark. 2005. Effects of small mammal cycles on productivity of boreal ducks. Wildlife Biology 11:3–11.

Camill, P., and J. S. Clark. 1998. Climate change disequilibrium of boreal permafrost peatlands caused by local processes. American Naturalist 151:207–222.

Carlson, M., J. Wells, and D. Roberts. 2009. The carbon the world forgot: conserving the capacity of Canada's boreal forest region to mitigate and adapt to climate change. Boreal Songbird Initiative and Canadian Boreal Initiative, Seattle, WA, and Ottawa, ON.

Chapin, F. S., T. V. Callaghan, Y. Bergeron, M. Fukuda, J. F. Johnstone, G. Juday, and S. A. Zimov. 2004. Global change and the boreal forest: thresholds, shifting states or gradual change? Ambio 33:361–365.

Christensen, T. K. 1999. Effects of cohort and individual variation in duckling body condition on survival and recruitment in the Common Eider *Somateria mollissima*. Journal of Avian Biology 30:302–308.

Corcoran, R. M., J. R. Lovvorn, M. R. Bertram, and M. T. Vivion. 2007. Lesser Scaup nest success and duckling survival on the Yukon Flats, Alaska. Journal of Wildlife Management 71:127–134.

Cox, R. R., Jr., M. A. Hanson, C. C. Roy, N. H. Euliss, Jr., D. H. Johnson, and M. G. Butler. 1998. Mallard duckling growth and survival in relation to aquatic invertebrates. Journal of Wildlife Management 62:124–133.

Devito, K., I. Creed, T. Gan, C. Mendoza, R. Petrone, U. Silins, and B. Smerdon. 2005. A framework for broad-scale classification of hydrologic response units on the boreal plain: is topography the last thing to consider? Hydrological Processes 19:1705–1714.

Devries, J. H., K. L. Guyn, R. G. Clark, M. G. Anderson, D. Caswell, S. K. Davis, D. G. McMaster, T. Sopuck, and D. Kay. 2004. Prairie Habitat Joint Venture (PHJV) waterfowl habitat goals update: Phase I. Prepared for Prairie Habitat Joint Venture Waterfowl Working Group.

Drever, C. R., G. Peterson, C. Messier, Y. Bergeron, and M. Flannigan. 2006. Can forest management based on natural disturbances maintain ecological resilience? Canadian Journal of Forest Research 36:2285–2299.

Ducks Unlimited Canada. 2010. Western Boreal Forest Conservation Plan. Internal document.

Elkie, P. C., and R. S. Rempel. 2001. Detecting scales of pattern in boreal forest landscapes. Forest Ecology and Management 147:253–261.

Fast, H., and F. Berkes. 1999. Climate change, northern subsistence, and land-based economies. Pp. 9–19 in D. Wall, M. M. R. Freeman, P. A. McCormack, M. Payne, E. E. Wein, and R. W. Wein (editors), Securing northern futures: developing research partners. Canadian Circumpolar Institute (CCI) Press, Edmonton, AB.

Fast, P. L. F., R. G. Clark, R. W. Brook, and J. E. Hines. 2004. Patterns of wetland use by brood-rearing Lesser Scaup in northern boreal forest of Canada. Waterbirds 27:177–182.

Findlay, C. S., and J. Bourdages. 2000. Response time of wetland biodiversity to road construction on adjacent lands. Conservation Biology 14:86–94.

Findlay, C. S., and J. Houlahan. 1997. Anthropogenic correlates of species richness in southeastern Ontario wetlands. Conservation Biology 11:1000–1009.

Foote, L., and N. Krogman. 2006. Wetlands in Canada's western boreal forest: agents of change. Forestry Chronicle 82:825–833.

Forman, R. T. T., and L. E. Alexander. 1998. Roads and their major ecological effects. Annual Review of Ecology and Systematics 29:207–231.

Greenwood, R. J., A. B. Sargeant, D. H. Johnson, L. M. Cowardin, and T. L. Shaffer. 1995. Factors associated with duck nest success in the prairie pothole region of Canada. Wildlife Monographs 128:3–57.

Guisan, A., and N. E. Zimmermann. 2000. Predictive habitat distribution models in ecology. Ecological Modelling 135:147–186.

Gurney, K. E., T. D. Williams, J. E. Smits, M. Wayland, S. Trudeau, and L. I. Bendell-Young. 2005. Impact of oil-sands based wetlands on the growth of Mallard (Anas platyrhynchos) ducklings. Environmental Toxicology and Chemistry 24:457–463.

Guyn, K. L., and R. G. Clark. 2000. Nesting effort of Northern Pintails in Alberta. The Condor 102:619–628.

Hill, M. R. J., R. T. Alisauskas, C. D. Ankney, and J. O. Leafloor. 2003. Influence of body size and condition on harvest and survival of juvenile Canada Geese from Akimiski Island, Nunavut. Journal of Wildlife Management 67:530–541.

Hobson, K. A., E. M. Bayne, and S. L. Van Wilgenburg. 2002. Large-scale conversion of forest to agriculture in the boreal plains of Saskatchewan. Conservation Biology 16:1530–1541.

Hobson, K. A., and J. Schieck. 1999. Changes in bird communities in boreal mixedwood forest: harvest and wildfire effects over 30 years. Ecological Applications 9:849–863.

Hodges, J. I., J. G. King, B. Conent, and H. A. Hanson. 1996. Aerial survey of waterbirds in Alaska: 1957–94: population trends and observer variability. National Biological Service Information and Technology Report 4.

Holling, C. S. 1973. Resilience and stability of ecological systems. Annual Review of Ecology and Systematics 4:1–23.

Holling, C. S. 1978. Adaptive environmental assessment and management. John Wiley, New York, NY.

Hornung, J. P., and A. L. Foote. 2006. Aquatic invertebrate responses to fish presence and vegetation complexity in western boreal wetlands, with implications for waterbird productivity. Wetlands 26:1–12.

Houlahan, J. E., and C. S. Findlay. 2004. Estimating the "critical" distance at which adjacent land-use degrades wetland water and sediment quality. Landscape Ecology 19:677–690

Hunter, M. L., Jr. 1993. Natural fire regimes as spatial models for managing boreal forests. Biological Conservation 65:115–120.

James, A. R. C., and A. K. Stuart-Smith. 2000. Distribution of caribou and wolves in relation to linear corridors. Journal of Wildlife Management 64:154–159.

King, J., and L. I. Bendell-Young. 2000. Toxicological significance of grit replacement times for juvenile Mallards. Journal of Wildlife Management 64:858–862.

Krapu, G. L., A. T. Klett, and D. G. Jorde. 1983. The effect of variable spring water conditions on Mallard reproduction. Auk 100:689–398.

Lake, B. C., J. Walker, and M. S. Lindberg. 2006. Survival of ducks banded in the boreal forest of Alaska. Journal of Wildlife Management 70:443–449.

Lancia, R. A., C. E. Braun, M. W. Collopy, R. D. Dueser, J. G. Kie, C. J. Martinka, J. D. Nicholas, T. D. Nudds, W. R. Porath, and N. G. Tilghman. 1996. ARM! for the future: adaptive resource management in the wildlife profession. Wildlife Society Bulletin 24:436–442.

Lee, K. N. 1993. Compass and gyroscope: integrating science and politics for the environment. Island Press, Washington, DC.

Lee, P., and S. Boutin 2006. Persistence and development of wide seismic lines in the western boreal plains of Canada. Journal of Environmental Management. 78:240–250

Lemelin, L.-V., L. Imbeau, M. Darveau, and D. Bordage. 2007. Local short-term effects of forest harvesting on breeding waterfowl and Common Loon in forest dominated landscapes of Quebec. Avian Conservation and Ecology. <http://www.ace-eco.org/vol2/iss2/art10/>.

Lindenmayer, D. B., and R. F. Noss. 2006. Salvage logging, ecosystem processes, and biodiversity conservation. Conservation Biology 20:949–958.

Lindenmayer, D. B., J. F. Franklin, and J. Fisher. 2006. General management principles and a checklist of strategies to guide forest biodiversity conservation. Biological Conservation 131:433–445.

Lynch, J. 1984. Escape from mediocrity: a new approach to American waterfowl hunting regulations. Waterfowl 35:5–13.

Margules, C. R., and R. L. Pressey. 2000. Systematic conservation planning. Nature 405:243–253.

National Ecological Assessment Team. 2006. Strategic habitat conservation. Final Report of the National Ecological Assessment Team. <http://www.fws .gov/nc-es/habreg/NEAT_FinalRpt.pdf>.

National Wetlands Working Group. 1997. The Canadian wetland classification system. 2nd ed. Wetlands Research Centre, University of Waterloo, Waterloo, ON.

Nielsen, S. E., E. M. Bayne, J. Schieck, J. Herbers, and S. Boutin. 2007. A new method to estimate species and biodiversity intactness using empirically derived reference conditions. Biological Conservation 137:403–414.

Niemela, J. 1999. Management in relation to disturbance in the boreal forest. Forest Ecology and Management 115:127–134.

Niemi, G., J. Hanowski, P. Helle, R. Howe, M. Monkkonen, L. Vernier, and D. Welsh. 1998. Ecological sustainability of birds in boreal forests. Conservation Ecology 2:17.

North American Waterfowl Management Plan, Plan Committee. 2004. North American Waterfowl Management Plan 2004. Strategic guidance: strengthening the biological foundation. Canadian Wildlife Service, U.S. Fish and Wildlife Service, Secretaria de Medio Ambiente y Reursos Naturales, Canada.

North American Waterfowl Management Plan, Plan Committee. 2007. North American Waterfowl Management Plan: continental progress assessment final report. Canadian Wildlife Service, U.S. Fish and Wildlife Service, Secretaria de Medio Ambiente y Reursos Naturales, Canada.

Pastor, J., S. Light, and L. Sovell. 1998. Sustainability and resilience in boreal regions: sources and consequences of variability. Conservation Ecology 2:16.

Paszkowski, C. A., and W. M. Tonn. 2000. Community concordance between the fish and aquatic birds of lakes in northern Alberta, Canada: the relative importance of environmental and biotic factors. Freshwater Biology 43:421–437.

Paszkowski, C. A., and W. M. Tonn. 2006. Foraging guilds of aquatic birds on productive boreal lakes: environmental relations and concordance patterns. Hydrobiologia 567:19–30.

Petrone, R. M., U. Silins, and K. J. Devito. 2007. Potential impacts of catchment microclimatic variability on pond evaporation in the western boreal plain, Alberta, Canada. Hydrological Processes 21:1391–1401.

Pierre, J. P., H. Bears, and C. A. Paszowksi. 2001. Effects of forest harvesting on nest predation in cavity-nesting waterfowl. Auk 118:224–230.

Price, J. S., B. A. Branfireun, J. M. Waddington , and K. J. Devito. 2005. Advances in Canadian wetland hydrology, 1999–2003. Hydrological Processes 19:201–214.

Rhymer, J. M. 1988. The effect of egg size variability on thermoregulation of Mallard (Anas platyrhynchos) offspring and its implications for survival. Oecologia 75:20–24.

Riordan, B., D. Verby la, and A. D. McGuíve, 2006, Shrinking ponds in subarctic Alaska based on 1950–2002 remotely sensed images. Journal of Geophysical Research 111:1–11.

Sargeant, A. B., and D. G. Raveling. 1992. Mortality during the breeding season. Pp. 396–422 in B. D. J. Batt, A. D. Afton, M. G. Anderson, C. D. Ankney, D. H. Johnson, J. A. Kadlec, and G. L. Krapu (editors), Ecology and management of breeding waterfowl. University of Minnesota Press, Minneapolis, MN.

Schindler, D. 1998. Sustaining aquatic ecosystems in boreal regions. Conservation Ecology 2:18.

Schmidt, J. H., E. J. Taylor, and E. A. Rexstad. 2006. Survival of Common Goldeneye ducklings in interior Alaska. Journal of Wildlife Management 70:792–798.

Schmiegelow, F. K. A., and M. Mönkkönen. 2002. Habitat loss and fragmentation in dynamic landscapes: avian perspectives from the boreal forest. Ecological Applications 12:375–389.

Schneider, R. 2000. Integrating protected areas and ecological forest management in Alberta's boreal forest.
Alberta Centre for Boreal Research, Edmonton, AB.

Schneider, R. R. 2002. Alternative futures: Alberta's
boreal forest at the crossroads. The Federation
of Alberta Naturalists, Alberta Center for Boreal
Research, Edmonton, AB.

Schneider, R. R., J. B. Stelfox, S. Boutin, and S. Wasel.
2003. Managing the cumulative impacts of land-
uses in the western Canadian sedimentary basin: a
modeling approach. Conservation Ecology 7:8.

Smith, L. C., Y. Sheng, G. M. MacDonald, and L. D.
Hinzman. 2005. Disappearing arctic lakes. Science
308:1429.

Szaro, R. C., G. Hensler, and G. H. Heinz. 1981.
Effects of chronic ingestion of no. 2 fuel oil on Mallard ducklings. Journal of Toxicology and Environmental Health, Part A 7:789–799

Trombulak, S. C., and C. A. Frissell. 2000. Review of
ecological effects of roads on terrestrial and aquatic
communities. Conservation Biology 14:18–30.

U.S. Fish and Wildlife Service. 2003. Strengthening
the biological foundations: update to the North
American Waterfowl Management Plan; first draft
for review by plan stakeholders. U.S. Department of
the Interior, Washington, DC.

U.S. Fish and Wildlife Service 2007. Waterfowl Popula-
tion Status, 2007. U.S. Department of the Interior,
Washington, DC.<http://www.fws.gov/migratorybirds/
NewReportsPublications/PopulationStatus/Waterfowl/
Status%20of%20waterfowl%202007.pdf>.

U.S. Fish and Wildlife Service and Canadian Wildlife
Service. 1987. Standard operating procedures for
aerial waterfowl breeding ground population and
habitat surveys in North America. Washington, DC.

Vitt, D. H. 1994. An overview of factors that influence
the development of Canadian peatlands. Memoirs
of the Entomological Society of Canada 169:7–20.

Walker, J., and M. S. Lindberg. 2005. Survival of scaup
ducklings in the boreal forest of Alaska. Journal of
Wildlife Management 69:592–600.

Walker, J., M. S. Lindberg, M. MacCluskie, M. J.
Petrula, and J. S. Sedinger. 2005. Nest survival of
scaup and other ducks in the boreal forest of Alaska.
Journal of Wildlife Management 69:582–591.

Walsh, K. A., D. R. Halliwell, J. E. Hines, M. A.
Fournier, A. Czarnecki, and M. F. Dahl. 2006.
Effects of water quality on habitat use by Lesser
Scaup (*Aythya affinis*) broods in the boreal
Northwest Territories, Canada. Hydrobiologia
567:101–111.

Walters, C. J. 1986. Adaptive management of renewable resources. McMillan, New York, NY.

Walters, C. J., and C. S. Holling. 1990. Large-scale
management experiments and learning by doing.
Ecology 71:2060–2068.

Zimpfer, N. L., W. E. Rhodes, E. D. Silverman, G. S.
Zimmerman, and M. D. Kone. 2009. Trends in
duck breeding populations, 1955–2009. Administra-
tive Report, 1 July 2009. U.S. Fish and Wildlife
Service, Division of Migratory Bird Management,
Laurel, MD.

Average breeding season populations (mean, 50 × 1,000) among biomes within the Traditional Survey area of North America (see Figure 3.3) by decade

Strata included in each biome: boreal (1–6, 12–18, 20–25, 36, 50, 77), prairies (26–35, 37–49, 75, 76), and tundra (8–11), 1960–2009 (mean, SD).

Species	Species Common Name (Scientific Name)	1960–1969 Boreal		1960–1969 Prairies		1960–1969 Tundra		1970–1979 Boreal		1970–1979 Prairies		1970–1979 Tundra	
ABDU	American Black Duck (*Anas rubripes*)	27	15	0	0	0	0	9	14	0	0	0	0
AGWT	American Green-winged Teal (*Anas crecca*)	792	220	412	233	52	26	744	223	790	235	94	47
AMWI	American Wigeon (*Anas americana*)	1,168	260	966	266	129	45	1,180	250	1,275	254	195	90
BWTE	Blue-winged Teal (*Anas discors*)	417	170	2,974	544	0	0	357	228	3,811	669	1	3
BUFF	Bufflehead (*Bucephala albeola*)	325	75	82	31	2	1	504	66	98	31	3	2
CANV	Canvasback (*Aythya valisineria*)	204	78	315	61	4	4	130	44	375	67	4	5
GADW	Gadwall (*Anas strepera*)	40	23	1,101	409	0	0	34	16	1,396	164	0	1
EIDI	King Eider (*Somateria spectabilis*) and Common Eider (*Somateria mollissima*)	0	0	0	0	48	14	0	0	0	0	23	6
GOLD	Common Goldeneye (*Bucephala clangula*) and Barrow's Goldeneye (*Bucephala islandica*)	338	138	38	16	30	16	280	80	61	24	32	7

Code	Species												
MERG	Common Merganser (*Mergus merganser*), Hooded Merganser (*Lophodytes cucullatus*), and Red-breasted Merganser (*Mergus serrator*)	266	44	12	13	10	9	253	102	20	14	6	7
SCAU	Lesser Scaup (*Aythya affinis*) and Greater Scaup (*Aythya marila*)	2,976	265	905	249	672	111	3,761	729	1,161	351	571	172
SCOT	Black Scoter (*Melanitta nigra*), Surf Scoter (*Melanitta perspicillata*), and White-winged Scoter (*Melanitta fusca*)	819	205	38	10	487	200	946	206	25	10	343	122
LTDU	Long-tailed Duck (*Clangula hyemalis*)	254	113	0	0	233	58	206	112	0	0	191	58
MALL	Mallard (*Anas platyrhynchos*)	1,978	436	4,019	819	78	28	1,951	412	5,274	886	92	39
NOPI	Northern Pintail (*Anas acuta*)	737	285	2,828	1,204	546	166	633	242	4,023	1,272	631	275
NSHO	Northern Shoveler (*Anas clypeata*)	262	128	1,306	435	10	8	218	91	1,554	349	35	22
REDH	Redhead (*Aythya americana*)	72	29	440	138	0	0	33	12	577	96	0	0
RNDU	Ring-necked Duck (*Aythya collaris*)	335	132	34	16	1	5	332	84	72	31	1	3
RUDU	Ruddy Duck (*Oxyura jamaicensis*)	35	19	266	79	0	0	17	10	315	135	0	0

APPENDIX 3.1 (*continued*)

APPENDIX 3.1 (CONTINUED)

Species	1980–1989						1990–1999						2000–2009					
	Boreal		Prairies		Tundra		Boreal		Prairies		Tundra		Boreal		Prairies		Tundra	
ABDU	6	7	0	0	0	0	28	16	0	1	0	0	28	17	0	0	0	0
AGWT	1,059	197	365	89	159	50	1,064	121	543	239	293	89	1,302	295	663	237	418	78
AMWI	1,277	308	664	301	262	76	1,236	241	597	119	369	86	1,198	265	523	110	454	81
BWTE	547	228	2,798	469	0	1	345	165	3,960	1,417	0	0	267	95	4,907	1,211	0	2
BUFF	492	95	105	20	6	4	578	81	176	41	2	1	614	125	225	42	1	1
CANV	176	49	296	67	21	18	203	46	352	121	16	13	185	60	376	122	23	13
GADW	57	28	1,236	101	1	1	73	18	2,106	661	4	5	62	14	2,361	355	3	2
EIDI	0	0	0	0	21	12	0	0	0	0	10	3	0	0	0	0	16	7
GOLD	338	103	58	19	29	12	476	228	75	30	14	6	507	104	105	42	11	5
MERG	343	67	39	18	14	5	482	121	30	14	19	7	613	178	26	16	31	16
SCAU	3,248	488	1,147	317	680	172	2,366	383	928	193	676	96	1,854	476	724	134	712	91
SCOT	958	135	14	7	489	129	516	118	7	7	350	58	527	179	1	1	370	48
LTDU	169	77	0	0	247	98	64	17	0	0	96	12	44	23	0	0	109	17
MALL	1,830	378	3,268	543	161	53	2,074	330	4,216	1,436	240	110	1,739	407	4,840	589	330	66
NOPI	647	210	1,516	597	669	206	401	98	1,332	569	640	108	471	160	1,366	456	776	121
NSHO	390	134	1,236	258	115	69	466	122	1,821	733	229	85	506	147	2,453	662	285	36
REDH	43	19	521	86	1	1	43	17	623	204	0	1	38	13	706	202	1	2
RNDU	452	74	116	40	9	12	756	218	96	46	5	4	872	186	109	22	8	6
RUDU	57	40	432	181	0	0	60	32	387	123	0	1	43	23	529	115	0	0

Breeding Distribution and Ecology
of Pacific Coast Surf Scoters

John Y. Takekawa, Susan W. De La Cruz, Matthew T. Wilson,
Eric C. Palm, Julie Yee, David R. Nysewander,
Joseph R. Evenson, John M. Eadie, Daniel Esler,
W. Sean Boyd, and David H. Ward

Abstract. Recent declines in sea duck populations have highlighted the need for additional basic research across the life cycle of these long-distance migratory birds. A lack of basic ecological information on Surf Scoters (*Melanitta perspicillata*), including the linkage between wintering and breeding areas and description of their nesting areas, is a major impediment to determining factors contributing to their decline. We marked 415 Surf Scoters with radio and satellite transmitters at four wintering areas along the Pacific coast to describe their breeding synchrony, sympatry, philopatry, and nesting areas selection in the northern boreal forest (NBF). Their primary breeding region was located in the western NBF centered on the Great Slave and Great Bear Lakes in the Northwest Territories, Canada, and their mean settling date (31 May) was remarkably synchronous (± 0.9 d). We developed a nearest-neighbor statistic C to examine nesting areas of individuals from different wintering areas and found that they were not clustered ($C = 0.031$, $P = 0.15$), but nests of eight individuals found in successive years were highly philopatric and within 1.2 ± 0.2 km of their previous locations. Finally, we compared nesting areas and random locations with stepwise selection in a second-order Akaike Information Criterion (AIC_c) analysis to identify the best models. Key landscape features included distance to snowline, elevation gradient, numbers of lakes, distance to treeline, and latitude. A nonparametric classification and regression tree (CART) showed that nesting areas were in an arc of habitat near snowline (<218 km), in lower elevation gradients (<14 m/km), and in areas with 3–6 lakes within 2×2 km. Climate change is predicted to have the greatest effects on more northern ecosystems, and NBF species like Surf Scoters with relatively inflexible breeding ecology may be adversely affected if they are unable to adapt quickly to rapidly changing conditions.

Key Words: Baja California, boreal forest, breeding, *Melanitta perspicillata*, migratory connectivity, nearest neighbor, Northwest Territories, Puget Sound, San Francisco Bay, San Quintin Bay, site fidelity, Strait of Georgia, Surf Scoter.

Takekawa, J. Y., S. W. De La Cruz, M. T. Wilson, E. C. Palm, J. Yee, D. R. Nysewander, J. R. Evenson, J. M. Eadie, D. Esler, W. S. Boyd, and D. H. Ward. 2011. Breeding distribution and ecology of Pacific coast Surf Scoters. Pp. 41–64 *in* J. V. Wells (editor). Boreal birds of North America: a hemispheric view of their conservation links and significance. Studies in Avian Biology (no. 41), University of California Press, Berkeley, CA.

nderstanding relationships between non-breeding and breeding populations is fundamental to our knowledge of migratory bird ecology. Although many studies have examined migratory populations during wintering or breeding periods, few cross-seasonal studies have been undertaken on individual birds across these primary life-cycle stages (Webster et al. 2002). The primary reason that such research has been limited is because of the great difficulty in relocating individuals at both ends of their migratory routes. However, the development of satellite telemetry over the past decade has made such studies feasible for larger migratory birds such as waterfowl.

Information on migratory connectivity has been exceedingly sparse for sea ducks (Tribe Mergini) in North America, most of which winter in the temperate coastal waters and breed in remote sub-Arctic and Arctic regions. One species, the Surf Scoter (*Melanitta perspicillata*), breeds from Labrador to Alaska and winters along the Pacific and Atlantic coasts (Savard et al. 1998). Surf Scoters are the only scoter endemic to North America, and little is known about their breeding distribution and abundance, because this species nests in very low densities across an extensive range (Savard et al. 1998).

Surf Scoters nest in the northern boreal forest (NBF), a region characterized by extremes in temperatures and precipitation, low plant species diversity, recurring disturbances such as fire, dramatic fluctuations in insect and vertebrate populations, and sparse human populations (Shugart et al. 1972, Chapin et al. 2006). Physical and biological processes in the NBF are shaped by low temperatures and permafrost soil; organisms residing in the NBF are adapted to low temperatures (Chapin et al. 2006). The NBF has been described as a region with relatively stable wetland habitats (Jessen 1981), but in the past three decades, many areas of the NBF in western North America have warmed more rapidly than any other region on earth (Serreze et al. 2000). Sea ducks are the most northerly nesting of the ducks (Goudie et al. 1994), and climate-induced change, including alteration of wetlands (Smol and Douglas 2007), has been predicted to have the greatest effect on northern ecosystems (Soja et al. 2007).

Migratory birds such as the Surf Scoter occupy the NBF for 3–4 months of the year. Their productivity is a compromise of competing migratory schedules (Drent et al. 2003), where a lack of resources early in the season is balanced against declining reproductive success with advancing date. Surf Scoters pair on the wintering grounds and begin laying eggs in early June. Males depart soon after egg laying while females tend broods (Savard et al. 1998). Their habitat preferences for nesting areas are not known, but the few nests that have been reported are well concealed under conifers at variable distances to open water (Savard et al. 1998).

Surf Scoters are counted annually during the breeding waterfowl surveys in May; however, scoter numbers are not well documented, because the extent and timing of the May surveys are targeted for Mallards (*Anas platyrhynchos*) and survey coverage is less extensive within the NBF than in other waterfowl breeding habitats (Smith 1995, Hodges et al. 1996). During the annual May breeding survey, the majority of Surf Scoters are found in the NBF strata. Nevertheless, long-term surveys indicate declining trends in the breeding population of Surf Scoters (Goudie et al. 1994, Hodges et al. 1996, Savard et al. 1998, Sea Duck Joint Venture 2001, USFWS 2002, Nysewander et al. 2004). The Surf Scoter population has been estimated at 536,000 breeding birds (Goudie et al. 1994), and over the past two decades, the population has decreased nearly 50% (Sea Duck Joint Venture 2001).

Coincident with the declines on the breeding grounds, the number of Surf Scoters wintering in Pacific coast estuaries also has declined over the past two decades (USFWS 2002, Nysewander et al. 2004). These wintering areas face a variety of anthropogenic threats, such as contaminant exposure, nonnative species invasions, aquaculture, physical habitat alteration, and disturbance (Nichols et al. 1986, Carlton et al. 1990, Cohen and Carlton 1995, Savard et al. 1998, Linville et al. 2002, Nysewander et al. 2004). A lack of basic ecological information on Surf Scoters, including the linkage between wintering and breeding areas and description of their nesting areas, is a major impediment to determining factors contributing to their decline.

In this study, we integrated results from Surf Scoters marked with satellite transmitters at four wintering areas along the Pacific coast. We documented the scope of their core breeding area, their nesting synchrony and site fidelity, and the level of connectivity of wintering and breeding areas. Finally, we examined landscape features of the NBF to determine if Surf Scoters selected

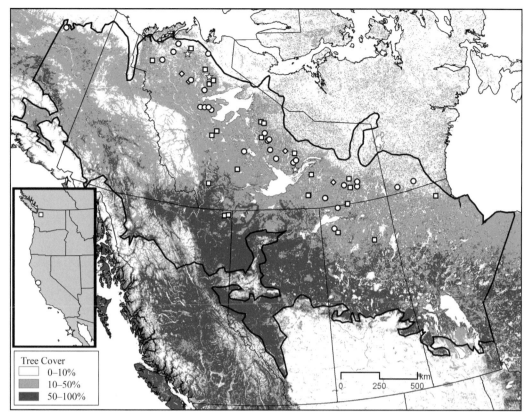

Figure 4.1. Outline of the breeding range extent for Surf Scoters in the northern boreal forest from eastern Alaska to the Ontario border, where shading indicates percent tree cover. Symbols show breeding locations for nesting Surf Scoters from different wintering areas, including San Quintin Bay, Baja California, Mexico (SQ, star); San Francisco Bay, California, USA (SF, circle); Puget Sound, Washington, USA (PS, square); and Strait of Georgia, British Columbia, Canada (SG, diamond). Inset shows four wintering areas where Surf Scoters were captured and marked.

specific areas for nesting, and how predicted climate-induced changes in the NBF may affect their breeding ecology.

METHODS

Study Area

The wintering range of Surf Scoters on the Pacific coast extends from Alaska to Baja California, Mexico (Savard et al. 1998). We compiled information from studies of Surf Scoters captured and marked at four wintering areas along the Pacific coast (Fig. 4.1). The wintering area name, coordinates, and approximate distance of their migration route to the breeding grounds (De La Cruz et al., 2009) included San Quintin Bay, Baja California, Mexico (SQ: 116.0°W, 30.4°N; 4,500 km), San Francisco Bay,

California, USA (SF: 122.4°W, 37.8°N; 3,750 km); Puget Sound, Washington, USA (PS: 122.4°W, 47.5°N; 2,000 km); and the Strait of Georgia, British Columbia, Canada (SG: 122.4°W, 49.3°N; 2,500 km). We examined landscape features within the breeding distribution of Surf Scoters in the NBF region of western Canada and eastern Alaska. We used tree density data from a digital coverage of the NBF (E. Butterworth, pers. comm.) georeferenced in ArcMap v. 9.2 (ESRI Inc., Redlands, CA) to delineate the southern extent of available nesting habitats and the treeline coverage for the northern extent. On the basis of our preliminary fieldwork and the satellite transmitter data in 2003, we defined longitude 148°W in eastern Alaska as the western boundary and the Manitoba–Ontario border (~95.2°W) as the eastern boundary of the breeding range for our analyses (Fig. 4.1).

Capture and Marking

We captured 415 Surf Scoters on their wintering grounds between November and March 2002–2006 with floating mistnets (Kaiser at al. 1995) or a net shot from a netgun on a fast-moving boat. Each bird was sexed and aged, banded, and measured. Selected individuals were abdominally implanted (Korschgen et al. 1996) with a platform-transmitter-terminal (PTT) satellite transmitter or very-high-frequency (VHF) radio transmitter with an external antenna. A study comparing different attachment methods (Iverson et al. 2006) concluded that this type of coelomic implant provides reliable, unbiased telemetry location data. Marked birds were released after a recovery period of at least two hours. VHF transmitters provided a signal every 1–2 sec for the life of the transmitter, while PTT transmitters had duty cycles to record location data for 6–8 h and off for 48–96 h, depending on the seasonal programming for the different project objectives.

Nesting Locations

Nesting locations of Surf Scoters were determined primarily from birds marked with satellite transmitters. Locations were obtained from the Argos data system, which estimated positions by calculating the Doppler-effect shift with receivers on National Oceanic and Atmospheric Administration (NOAA) polar-orbiting weather satellites. In 2003, 2005, and 2006, nesting locations were supplemented from extensive aerial searches of Surf Scoters marked with VHF transmitters. Observers used 1–2 receivers to listen and locate marked individuals as they flew in fixed-wing aircraft (Cessna 185 or 206) from Yellowknife (114.3°W, 62.5°N) outfitted with dual four-element Yagi antennas on their wing struts. Telemetry flights were flown in a grid pattern over most of the NBF in the Northwest Territories from Inuvik to the Alberta border and from the MacKenzie River in the west to treeline in the east. The initial, extensive aerial searches were conducted at higher elevations (>1,500 m) with an estimated detection range of 50 km for a total strip width of 100 km. Subsequent intensive searches were conducted at lower elevation (<800 m), focused on a core area (300 km × 900 km) between the Great Slave and Great Bear Lakes.

For this analysis, we compiled the location data from all of the PTT-marked Surf Scoters and removed those with Argos location quality (LQ) classes of 0 (<1,500 m error), A, B, or Z (error not estimable). Although field comparison studies indicate that accurate locations are obtained for a large proportion of locations in these lower LQ classes (see Miller et al. 2005), we used a conservative approach to describe the breeding distribution by analyzing only those locations with Argos LQ codes of 1 (<1,000 m), 2 (<350 m), or 3 (<150 m). We discarded locations that were separated by less than an hour, since an hour was sufficient time for a scoter to cross its breeding home range, thereby reducing potential autocorrelation (White and Garrott 1990).

The center of the nesting area for each bird was determined in ArcMap with the "mean center" tool in ArcToolbox. The distance from each location to the mean center was calculated with Hawth's Tools (Beyer 2004), and outliers more than two standard deviations from the mean were deleted (<4% of the data). The remaining data were used to estimate nest site locations for habitat analyses. Five Surf Scoters had bimodal distributions that may have represented two nesting attempts in separate areas; for these cases, we used the first cluster of locations for the analyses. Locations were placed into 2 × 2 km or 4 × 4 km grids (UTM Zone 10, NAD83; ArcMap Fishnet Tool; ESRI 1996) that covered the breeding range for our habitat analyses. We used these two grid sizes to examine landscape features at different scales and to make use of available satellite datasets.

We defined the primary breeding period as extending from 15 May to 1 July on the basis of our preliminary search data and reported nest timing for Surf Scoters (Savard and Lamothe 1991, Savard et al. 1998). Although nests were not confirmed for the majority of the birds, we followed a similar approach to Ely et al. (2006, 2007) to identify where repeated locations in a small area indicated likely breeding activity. Surf Scoters are single-brooded and have an estimated incubation period of 28–30 days (Savard et al. 1998), so we defined the settling date (SETD) for a bird as the first day that it arrived to a nesting area (typically <20 km^2) where an individual was located for ≥25 days. Since Surf Scoters have one of the shortest seasonal pair bonds among waterfowl and males leave females within three weeks after arrival on breeding lakes (Savard et al. 1998), we only analyzed data from females. We tested SETD differences among birds from different wintering areas with analysis of variance and presented 95% confidence intervals.

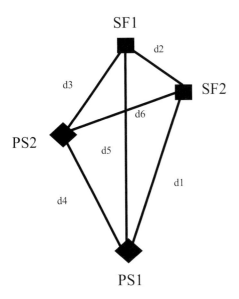

SF1

SF2

PS2

PS1

d1, d2, d3, d4, d5, d6

$$\bar{d} = (d1 + d2 + d3 + d4 + d5 + d6)/6$$
$$C_1 = f(SF1) = 1/(d2 + \bar{d})$$
$$C_2 = f(SF2) = 1/(d2 + \bar{d})$$
$$C_3 = f(PS1) = 1/(d4 + \bar{d})$$
$$C_4 = f(PS2) = 0/(d3 + \bar{d})$$
$$C = \sum_{n=1}^{4} Ci$$

Figure 4.2. Depiction of method used to calculate a nearest-neighbor cluster statistic C to examine differences in breeding locations for Surf Scoter areas wintering at four areas along the Pacific coast. Calculations were conducted for all pairs of locations, and statistical significance was determined with Monte Carlo simulations for 10,000 iterations under a null hypothesis of random mixing.

Nearest-Neighbor Analysis

We adapted the two-cluster, nearest-neighbor approach of Schilling (1986) and applied to biological problems by Rosing et al. (1998) along with the scale-independent approach and weighting method of Day et al. (1989; see also Cuzick and Edwards, 1990) to develop our analysis. The null hypothesis was that the locations of marked Surf Scoters from different wintering areas were thoroughly mixed and distributed randomly, while our alternative hypothesis was that they were found in non-random clusters. For each scoter, i, we determined the distance, d_i, to the nearest scoter from any wintering area, and calculated a cluster statistic function, $C_i = \delta_i/(d_i + \bar{d})$, where δ_i is an indicator function of whether the nearest scoter is of the same wintering area as scoter i, and $\bar{d}$ is the average distance between all pairs of locations, independent of wintering area affiliation. The nearest neighbor cluster statistic C was set equal to the sum of all C_{-i} (Fig. 4.2).

We determined statistical significance with simulation tests by reassigning the wintering areas randomly to the configuration of locations while preserving the original sample sizes and recalculating C. We conducted a Monte Carlo simulation with 10,000 iterations to generate the distribution of C under the null hypothesis of random mixing. After determining the overall result among the four wintering areas, we repeated the Monte Carlo simulation of 10,000 iterations for each pair of wintering areas to examine spatial separation.

Environmental Data

Boreal forest habitats have been characterized at different scales from general classification systems (Rempel et al. 1997), linking structure (environmental and biotic variables) and function (waterfowl use and productivity) to ground surveys relating wetland features (e.g., depth, area, perimeter, vegetation, amphipods) to brood-rearing use (Fast et al. 2004). We used remote-sensing data sets and characterized landscape features at a scale appropriate to describe a scoter's nesting area, determined from satellite telemetry locations with a rough accuracy of about 1 km. The landscape features are defined below and included: snow-free date, snow-free date difference, latitude, distance to treeline, distance to snowline on settling date, elevation, elevation gradient, number of lakes, water cover, ground cover, and tree cover. Elevation gradient and cover classes were estimated at two grid scales (2 × 2 km, 4 × 4 km) similar to the scale of the satellite locations (1 km) to reduce potential cross-scale correlations (Battin and Lawler 2006). We used Hawth's Tools (Beyer 2004) to generate a set of random points, equal in number to the nesting birds found in a particular year (2003–2006), and to compare general NBF landscape features with the scoter nesting areas in an unconstrained design (Battin and Lawler 2006). Habitats unsuitable for nesting, such as mountain ranges devoid of trees, and very large lakes (i.e., Great Bear, Great Slave, Athabasca) were excluded from the analysis.

We estimated the snow-free date (SFDA), or the Julian Date when an area became snow-free, from the Interactive Multisensor Snow and Ice Mapping System (IMS), Daily Northern Hemisphere Snow and Ice Analysis coverage (NOAA/NESDIS/OSDPD/SSD 2006). A grid file was obtained for each day between 15 May and 1 July from 2003 to 2006 that was coded as 0 = no data, 1 = open water, 2 = snow-free land, 3 = ocean or lake ice, or 4 = snow-covered land. The files were converted to raster format and displayed in ArcMap. Although the coverage was updated daily, the IMS is manually derived based on visual imagery, and persistent cloud cover may have resulted in large changes in snow cover between some updates. Snow cover varied greatly among years, so we used the same number of random points as nesting birds in each year to make the data set balanced. Thus, there were 7 random points for 2003, 3 for 2004, 26 for 2005, and 17 for 2006, and a random date was selected from the breeding period (15 May to 1 July) and assigned to that point.

We estimated snow-free date difference (SFDD) as the number of days between the settling date for a scoter and the snowmelt date at its nesting area. Negative values indicated by how many days the settling date preceded the snowmelt, zero indicated that the bird arrived on the date the area became snow-free, and positive values indicated the number of days after snowmelt that the bird arrived.

Distance to snowline (DSNO) was estimated as minimum distance (km) from the mean location of a nesting bird to the snowline (the boundary line coverage rather than individual pixels estimated from the IMS) on that bird's settling date as the snowline retreated to the northeast.

We estimated the distance to treeline (DTRE) as the minimum distance from the nesting location to the nearest edge of the treeline in Canada (treeline coverage, Ducks Unlimited Canada). The coverage did not extend into eastern Alaska, so here we estimated the treeline edge by determining the percent of tree cover (see vegetation cover below) below 5%.

Latitude (LATD) was recorded as decimal degrees for random points and the estimated center point of the nesting areas.

Elevation (ELEV; meters) and elevation gradient (ELG2, ELG4) at 2 × 2 km and 4 × 4 km scales were estimated from topographic coverages

converted from program Mapsource (Garmin International, Inc., Topo Canada, v. 2, Olathe, KS). Topographic maps were obtained for each location and registered with three control points. Program FastStone Capture was used to save these maps as image files, which were then reprojected for analyses to NAD 83, Zone 10 (ArcCatalog). For elevation gradient, we calculated the difference between the minimum and maximum elevation (m) and divided it by the horizontal distance (km) across a grid square at a particular scale.

The number of lakes (NLK2, NLK4) within 2 × 2 km or 4 × 4 km grids was estimated from the topographic coverages. A lake was included if any part of it was found within a particular grid. Water cover (WCV2, WCV4) was determined at 2 × 2 km and 4 × 4 km scales from the topographic coverages. Water cover estimated the total percentage of a grid covered by water, unlike the NLK2 and NLK4 variables that determined whether Surf Scoters were found in areas with a few large lakes or a small number of lakes. For each bird and random point, we used the screen capture (FastStone Capture) and grayscale conversion (Scion Image) programs. The spatial scale of the images was standardized (Tool "Set Scale") and pixels with water were selected (Tool "Density Slice") by their grayscale values to estimate the extent of lakes and streams (Tool "Measure").

Vegetation cover was determined in 2 × 2 km and 4 × 4 km grids from satellite data obtained by the Moderate-resolution Imaging Spectroradiometer (MODIS) instrument and the vegetation continuous field coverage (Hansen et al. 2001). The MODIS satellite data were projected in WGS 84 with a resolution of 500 m, and we used the vegetation continuous field data (Hansen et al. 2001) to obtain estimates of vegetation cover from 31 October 2000 to 9 December 2001 (Global Land Cover Facility, College Park, MD, http://www.landcover.org). We selected grid cells with the Thematic Raster Summary Tool (Beyer 2004) to determine the percent coverage of each cover type. Each pixel included a value for percent ground cover (GCV2, GCV4), tree cover (TCV2, TCV4), and bare ground (BAR2, BAR4). The cover types comprised a composition that added to 1; therefore, we reported the means but excluded percent cover of bare ground (BAR2, BAR4) in analyses since it was dependent on the other variables.

TABLE 4.1

Landscape features (mean ± SE) for Surf Scoters (n = 53) nesting in the northern boreal forest (NBF) from different wintering areas

Detailed explanation of variables is provided in Methods.

Variable	San Quintin Bay, Baja California Norte, Mexico	San Francisco Bay, California, USA	Puget Sound, Washington, USA	Strait of Georgia, British Columbia, Canada
Snow-free date (Julian Date)	146.0 ± 2.0	145.5 ± 2.5	148 ± 2.7	148.5 ± 3.8
Snow-free date difference	14.3 ± 3.3	4.2 ± 2.3	2.2 ± 2.4	2.8 ± 2.3
Latitude in decimal degrees	66.1 ± 1.9	64.0 ± 0.6	63.3 ± 0.7	64.8 ± 1.5
Distance to treeline	116 ± 29	174 ± 20	224 ± 35	169 ± 28
Distance to snowline	56 ± 33	77 ± 16	75 ± 14	34 ± 10
Elevation	246 ± 25	312 ± 26	349 ± 26	329 ± 53
Variables at two grid scales (2 × 2 and 4 × 4 km)				
Elevation gradient (2 × 2 km)	7.1 ± 3.5	5.0 ± 1.2	6.9 ± 1.6	7.1 ± 4.1
Elevation gradient (4 × 4 km)	4.1 ± 2.1	7.8 ± 2.1	5.6 ± 1.3	5.8 ± 2.2
Number of lakes (2 × 2 km)	6.7 ± 2.3	6.7 ± 0.9	4.0 ± 0.5	4.8 ± 0.8
Number of lakes (4 × 4 km)	23.3 ± 2.6	24.4 ± 8.3	14.0 ± 3.6	13.0 ± 2.8
Water cover (2 × 2 km)	0.283 ± 0.145	0.165 ± 0.020	0.190 ± 0.037	0.133 ± 0.034
Water cover (4 × 4 km)	0.180 ± 0.075	0.150 ± 0.014	0.147 ± 0.020	0.163 ± 0.040
Bare ground (2 × 2 km)	0.050 ± 0.010	0.082 ± 0.011	0.066 ± 0.012	0.113 ± 0.029
Bare ground (4 × 4 km)	0.073 ± 0.015	0.090 ± 0.009	0.065 ± 0.011	0.095 ± 0.015
Ground cover (2 × 2 km)	0.520 ± 0.122	0.572 ± 0.017	0.542 ± 0.031	0.628 ± 0.052
Ground cover (4 × 4 km)	0.597 ± 0.066	0.589 ± 0.016	0.567 ± 0.017	0.603 ± 0.031
Tree cover (2 × 2 km)	0.147 ± 0.015	0.179 ± 0.015	0.203 ± 0.023	0.133 ± 0.040
Tree cover (4 × 4 km)	0.153 ± 0.015	0.172 ± 0.014	0.222 ± 0.023	0.143 ± 0.027

Nesting Area Selection

We used logistic regression to model the selection of nesting areas on the basis of characteristics associated with the timing and location of breeding (Table 4.1; Manly et al. 2002). We applied second-order Akaike's Information Criterion (AIC_c; Burnham and Anderson 2002) to select the best model from a series of candidate logistic regression models comparing characteristics of scoter nesting areas with randomly selected points. Small sample sizes of birds in some wintering areas limited our ability to test differences among years; therefore, we pooled the samples across years. AIC values and Akaike weights were calculated for candidate models under logistic regression (Burnham and Anderson 2002, SAS Institute 2004).

We used a second-order AIC: $AIC_c = -2(\text{log-likelihood}) + 2KN/(N - K - 1)$, where K is the number of fitted parameters including variance and N is the sample size (Anderson et al. 2000; Burnham and Anderson 2002). We considered the model with the smallest AIC_c to be the most parsimonious (Anderson et al. 2000; Burnham and Anderson 2002). We calculated the AIC_c differences between the best model and the other candidate models ($\Delta AIC_{ci} = AIC_{ci} - \text{minimum } AIC_c$). Akaike weights ($w_i = \exp[-\Delta AIC_{ci}/2]/\Sigma \exp[-\Delta AIC_{ci}/2]$) were calculated to assess the evidence that the selected model was the best Kullback–Leibler model (Anderson et al. 2000; Burnham and Anderson 2008).

The number of possible models built from combinations of 17 variables ($n = 2^{17} - 1$) exceeded that which could be reasonably examined to determine their relative ranking. However, estimating AIC_c for a smaller subset of the possible variable combinations could exclude the best models (Steidl 2006), and it was not clear that grouping certain variables would be appropriate. Therefore, we conducted model selection by AIC in multiple stages. Our initial set of models was selected *a priori* to any analyses and included single-effect models and the full-term model containing all 17 variables. We hypothesized that any effect of latitude could differ among Surf Scoters from different wintering populations (WPOP), so we included WPOP and LATD*WPOP terms in the full-term model. If the full-term model fit best, then this indicated a better model existed based on combinations of variables. We then used a stepwise approach as recommended by Steidl (2006) to find the models with combinations of variables that produced the lowest AIC_c value.

We conducted both a backward stepwise analysis, removing a single variable at a time from the full-term model, and a forward stepwise analysis, entering a single variable at a time to the null (intercept-only) model. We used AIC_c as a criterion for each forward or backward step by selecting the variable in which its addition or removal contributed to the greatest reduction in the AIC_c, until no single addition or removal would further reduce the AIC_c. Since different models can result from forward and backward stepwise selection, we used AIC_c to select the better of the two models. We included any model with a ΔAIC_c within 2 units of the best model and reported the evidence ratio to compare the relative likelihood of probability between two models (Burnham and Anderson 2002). We calculated odds ratios (OR) for variables in the best model to determine if locations with higher values of these variables were more (OR > 1) or less likely (OR < 1) to become nesting area locations.

We then applied a classification and regression tree (CART) to confirm the direction and significance of effects in the best model (Stephens et al. 2005) and to examine cross-scale correlation (Battin and Lawler 2006). CART models are in the form of nonparametric, dichotomous keys (Brieman et al. 1984, De'ath and Fabricus 2000, Maisonneuve et al. 2006). CART algorithms partitioned the locations into subsets by recursively splitting explanatory variables into high and low categories which significantly contribute to the prediction of Surf Scoter nesting areas. The recursive partitioning of the data into subcategories within categories enabled our analysis to naturally explore and identify interaction effects, which would have been cumbersome with the logistic regression approach. CART was estimated from the "modeltools" and "party" packages for R software at the 0.10 significance criterion (R Development Core Team 2008; Hothorn et al. 2006, 2008).

RESULTS

Total Marked

From 2003 to 2006, 415 Surf Scoters were captured and radio-marked on the Pacific coast (Table 4.2) during the nonbreeding season, including 313 VHF-marked and 102 PTT-marked individuals. When we censored the males, the total sample was reduced by 66 individuals (16%). Of the remaining 349 marked Surf Scoters, only 55 females were found at inland locations where they stayed long enough to confirm their nesting locations. Eight of the 55 birds were located in two consecutive years (2005–2006), but we only used the first year in our analyses, so all samples were independent and based on different individuals. Finally, we located nests for three Surf Scoters, one PTT-marked bird and two VHF-marked birds, but we lacked repeated locations for the VHF-marked birds to estimate environmental conditions when they settled at nesting areas. Thus, we used a sample size of 53 Surf Scoters for our analyses.

Settling Date

We used a conservative approach to estimate SETD for nesting Surf Scoters by (1) restricting our analyses to Argos location quality classes 1–3 and (2) limiting the birds to those that were repeatedly located at a site for at least 25 days. Our analysis indicated the mean SETD ($\pm$SE) was Julian Date 151 $\pm$ 0.9 d, or 31 May. Mean SETD varied ($F_{3,52} = 2.70$, $P = 0.056$) by wintering area, ranging from PS: 149.7 $\pm$ 1.1 ($n = 21$); SF: 149.8 $\pm$ 1.4 ($n = 25$); SG: 151.3 $\pm$ 3.2 ($n = 4$); and SQ: 160.3 $\pm$ 4.7 ($n = 3$). The 95% confidence intervals (SQ: 153.2–167.4; SG: 145.1–157.4; SF: 147.3–152.2; and PS: 149.7–152.3) indicated that SETD was later for the southernmost wintering area (SQ), but was overlapping for the other wintering areas.

TABLE 4.2

Summary of Surf Scoters marked with very high frequency (VHF) or platform transmitter terminal (PTT)
satellite transmitters in 2003–2006 from four different wintering areas

Numbers in parentheses denote the total number of birds used for breeding ground analyses (see Methods).

Year	San Quintin Bay, Baja California Norte, Mexico		San Francisco Bay, California, USA[a]		Puget Sound, Washington, USA		Strait of Georgia, British Columbia, Canada	
	VHF	PTT	VHF	PTT	VHF	PTT	VHF	PTT
2003	0	0	33	8 (7)	0	0	0	0
2004	0	0	0	0	25	14 (3)	0	0
2005	14	5 (1)	87	11 (9)	35	22 (12)	77	8 (4)
2006	17	5 (2)	40	12 (9)	30	17 (6)	0	0
Total[b]	31	10 (3)	127	31 (25)	78	53 (21)	77	8 (4)

[a] The nests of three San Francisco Bay birds (2 VHF, 1 PTT) were located in ground searches, but the two VHF birds were not included in analyses because their settling dates were unknown.

[b] Eight birds marked in 2005 and also located in 2006 are not included twice.

Breeding Distribution

Surf Scoters from Pacific coast wintering areas used a relatively narrow, core breeding area from Lake Athabasca northwest to the Anderson River near the Arctic Ocean, with the distribution centered on the Great Slave and Great Bear Lakes in the Northwest Territories (Fig. 4.1). Surf Scoters were distributed between 57.7°N and 69.0°N latitude, and 147.2°W and 96.5°W longitude. Nesting locations for eight birds from two different wintering areas (SF = 5, PS = 3) were located in two consecutive years and showed remarkable nesting area fidelity (Table 4.3; Fig. 4.3). The center of their nesting areas was located within 1.3 km ± 0.2 km of each other in subsequent years, but

TABLE 4.3

Remarkable nesting area fidelity of Pacific coast Surf Scoters found in the northern boreal forest in two consecutive years

Wintering area, settling date (Julian Date), longitude and latitude (decimal degrees),
and mean distance between nesting areas (km) are reported for each individual.

Bird[a]	Wintering Area	Year 1			Year 2			Distance (km)
		Settling Date	Longitude (°W)	Latitude (°N)	Settling Date	Longitude (°W)	Latitude (°N)	
40843	PS	146	123.161	64.169	167	123.162	64.165	0.52
43888	PS	145	119.623	62.360	145	119.622	62.355	0.63
43892	PS	146	116.584	64.184	158	116.596	64.179	0.71
53978	PS	149	106.645	60.021	152	106.620	60.024	1.47
53980	PS	146	111.058	60.781	144	111.116	60.778	3.17
55912	SF	159	109.149	60.521	143	109.154	60.521	0.33
55914	SF	144	147.198	67.572	154	147.201	67.574	0.34
55919	SF	158	105.202	60.899	149	105.153	60.888	2.94
Mean	—	149	—	—	152	—	—	1.26

[a] Located in 2005 and 2006, except for 40843 and 43892, located in 2006 and 2007.

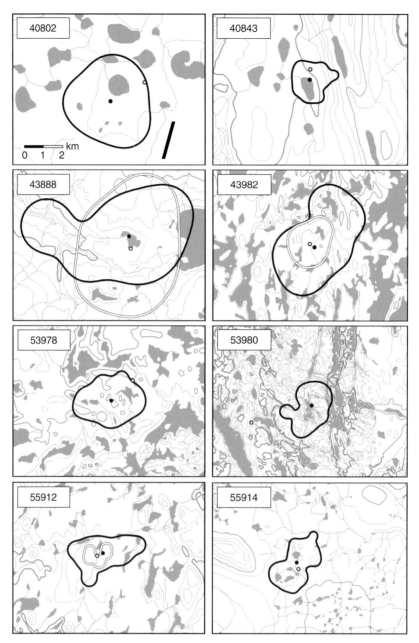

Figure 4.3. Remarkable breeding site philopatry exhibited by Surf Scoters located at the same nesting areas in two consecutive years. The center of the home range (circles) and 95% fixed kernel home ranges (lines) are depicted for 2005 (closed circle or line) and 2006 (open circle or line). Individuals shown are from Puget Sound (40802, 40843, 43888, 43982, 53978, 53980) and San Francisco Bay (55912, 55914) wintering areas, but sample sizes were only adequate to create home ranges for three individuals (43888, 43982, 55912) in 2006.

TABLE 4.4

Nearest-neighbor analysis for Surf Scoters wintering at four areas on the Pacific coast

Table values include sample size (*n*), the generated C-statistic, expectation of C,
standard deviation of C, and probability on the basis of 10,000 Monte Carlo iterations.

Comparison	*n*	*C*	E(*C*)	SD(*C*)	Prob.
Overall	53	0.031	0.025	0.0052	0.1496
San Quintin Bay, Strait of Georgia	7	0.003	0.004	0.0022	0.6612
San Quintin Bay, San Francisco Bay	28	0.027	0.027	0.0020	0.3434
San Quintin Bay, Strait of Georgia	24	0.019	0.023	0.0019	0.9633
Strait of Georgia, San Francisco Bay	29	0.029	0.026	0.0023	0.1714
Strait of Georgia, Puget Sound	25	0.026	0.022	0.0024	0.0764
San Francisco Bay, Puget Sound	46	0.032	0.028	0.0052	0.1873

since we were unable to locate the exact nest location, we could not confirm if they used the same nest site.

Nearest-Neighbor Analysis

We were unable to reject the overall null hypothesis that breeding Surf Scoters from different Pacific coast wintering areas were thoroughly mixed and distributed randomly (Table 4.4). Surprisingly, the strongest evidence (*P* = 0.0764) for clustering of Surf Scoters on the breeding grounds was among those individuals from PS and SG, separated by only 200 km in the winter. We were unable to reject random mixing in pairwise comparisons of the other wintering areas.

Nesting Area Selection

We were able to locate three nests of Surf Scoters by landing and searching the vicinity of locations in 2005–2006 (Fig. 4.4A–D). In general, nesting site vegetation was characterized by sparse black spruce (*Picea mariana*) trees (>4 m in height, 90% cover above the nest) and ground cover in lakeshore bogs comprised primarily of Labrador tea (*Ledum groenlandicum*), cloudberry (*Rubus chamaemorus*), mountain cranberry (*Vaccinium vitis-idaea*), and lichen. Two of the nesting sites were found in granitic or metamorphic rock outcroppings with sparse (0–5%) soil cover. The nests were found near

oligotrophic shallow lakes, but not where other waterfowl were observed.

Comparison among models of nesting area variables indicated that the backward selection model with ten variables most closely fit the data (Akaike weight = 81%; Table 4.5). The forward stepwise procedure resulted in an eight-variable model with a much poorer fit to the data (ΔAIC_c = 2.9, Akaike weight = 19%). In the forward selection model, DSNO entered first, followed by ELG2 and ELG4, with strong evidence in favor of entering versus not entering each variable (ΔAIC_c > 5.9; evidence ratio > 19). The variables NLK4, NLK2, and LATD entered next under weaker evidence (ΔAIC_c < 0.76; evidence ratio < 1.5), followed by DTRE (ΔAIC_c = 4.8; evidence ratio = 11.3), and finally TCV4 (ΔAIC_c = 0.88; evidence ratio = 1.6). Both the backward and forward selection models fit the data better than the full-term model or the null model (ΔAIC_c > 26; evidence ratio = 100,000) and contained seven variables in common: DSNO, ELG2, ELG4, DTRE, LATD, NLK2, and NLK4. Odds ratios for these variables based on model averaging suggested that nesting Surf Scoters tended to settle nearer to snowline (OR = 0.983; 95% CI = 0.971–0.995; Fig. 4.5), in areas with less elevation gradient at 2 × 2 km (OR = 0.912; 95% CI = 0.857–0.971) and 4 × 4 km (OR = 0.960; 95% CI = 0.909–1.013) scales, in areas with fewer lakes at 2 × 2 km (OR = 0.784; 95% CI = 0.639–0.964) but more lakes at 4 × 4 km (OR = 1.246; 95% CI = 1.057–1.470), farther from treeline (OR = 1.013; 95%

Figure 4.4. Surf Scoters nesting within the northern boreal forest: (A) aerial view of nesting site, (B) ground view of nesting site, (C) female scoter on a nest, and (D) clutch of Surf Scoters eggs (11 June). USGS photo credit: E. Palm.

TABLE 4.5
Ranking of candidate Akaike Information Criterion (AIC) models
to describe nesting areas of Surf Scoters breeding in the northern boreal forest

Backward and forward stepwise methods were used to select the two best models from the numerous combinations of the 17 variables that are presented with the full-term model (all variables) and null model (intercept only).

Model	N	K	−2 Log-likelihood	AIC$_c$	ΔAIC$_c$	Akaike Weight (%)
LATD, DTRE, DSNO, ELEV, ELG2, ELG4, NLK2, NLK4, WCV2, GCV2[a]	106	11	54.1	78.9	0.0	81
LATD, DTRE, DSNO, ELG2, ELG4, NLK2, NLK4, TCV4[b]	106	9	61.9	81.8	2.9	19
SFDA, SFDD, LATD, WPOP, LATD*WPOP, DTRE, DSNO, ELEV, ELG2, ELG4, NLK2, NLK4, WCV2, WCV4, GCV2, GCV4, TCV2, TCV4[c]	106	23	46.0	105.5	26.6	0
Null Model (intercept only)[d]	106	0	146.9	146.9	68.1	0

NOTE: Variables include snow-free date (Julian Date: SFDA); snow-free date difference (SFDD); latitude in decimal degrees (LATD); wintering population (WPOP); distance to treeline (m; DTRE); distance to snowline (m; DSNO); elevation (m; ELEV); and variables at two grid scales (2 × 2 km and 4 × 4 km), including elevation gradient (ELG2, ELG4), number of lakes (NLK2, NLK4), water cover (WCV2, WCV4), ground cover (GCV2, GCV4), and tree cover (TCV2, TCV4). Detailed explanation of variables is provided in Methods.

[a] Backward selection model includes intercept and 10 variables.

[b] Forward selection model includes intercept and 8 variables.

[c] Full model includes intercept, 17 variables, and LATD*WPOP interaction.

[d] In null model, k is zero since the intercept for the null model is known with equal sample sizes of nesting areas and random locations.

CI = 1.003–1.023), and at higher latitudes (OR = 1.500; 95% CI = 1.087–2.072).

The CART analysis (Fig. 4.6) split variables in a categorical tree that included elements from both the forward and backward selection models, consistent with the effects of DSNO, NLK4, ELG4, and ELG2 determined by the logistic regression. DSNO was the first significant predictor, with nesting areas comprising 2.9% of locations (1/34) at >218 km from snowline compared with 72.2% (52/72) at ≤218 km (left branch). Two variables at the larger (4 × 4 km) scale, NLK4 and ELG4, were the next most important predictors depending on distance to snowline. For locations >218 km from snowline, nesting sites comprised 9.1% (1/11) of locations with >7 lakes, compared with 0% (0/23) where there were ≤7 lakes. For locations ≤218 km from snowline, nesting locations comprised 33.3% (3/9) when elevation gradient was >14.1 m/km at the larger scale, compared with 77.8% (49/63) when ≤14.1 m/km (middle branch). For these latter locations, nesting areas comprised 89.7% (26/29) of locations at the smaller scale (2 × 2 km) where elevation gradient was ≤3.5 m/km, compared with 67.6% (23/34)

of locations where elevation gradient was higher. The CART analysis detected positive and negative effects associated with number of lakes at the small scale, depending on the elevation gradient. For low elevation gradients, nesting locations comprised 95.0% of locations with ≤6 lakes (19/20), compared to 77.8% where there were more lakes (7/9). In contrast, for higher elevation gradients, nesting locations comprised only 37.5% of locations with ≤6 lakes (6/16), compared to 94.4% where there were more lakes (17/18).

DISCUSSION

We integrated data sets from four different wintering areas along the Pacific coast to delineate the western breeding range for our sample of radio-marked Surf Scoters. Rather than finding wide temporal and spatial variation that might be expected for a species with a large breeding range in an ecosystem characterized by extreme variation (Chapin et al. 2006), we found that Surf Scoters settled in a relatively narrow arc of habitat parallel to the receding snowline. Also, the timing of their breeding was synchronous, as most

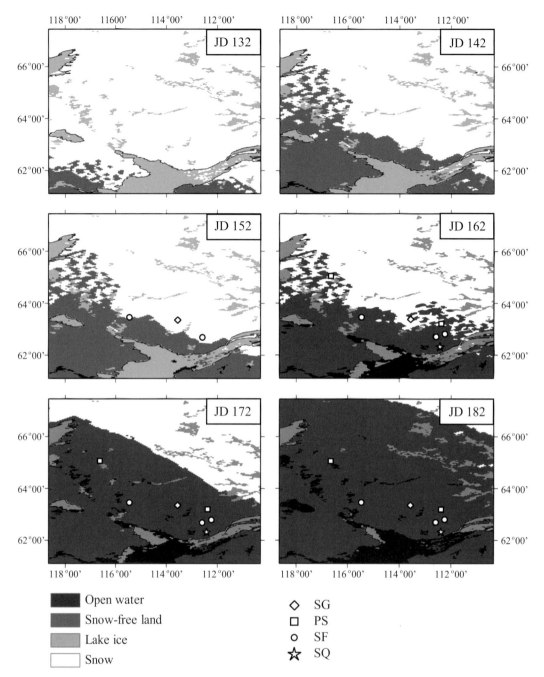

Figure 4.5. Seasonal variation (10-day intervals) in availability of habitats for Pacific coast Surf Scoters breeding in the northern boreal forest, where Julian Date (JD) 152 = 1 June. Snow-melt progression was determined with daily satellite images (4 × 4 km pixels) from the Interactive Multisensor Snow and Ice Mapping System (IMS), Daily Northern Hemisphere Snow and Ice Analysis coverage (NOAA/NESDIS/OSDPD/SSD 2006).

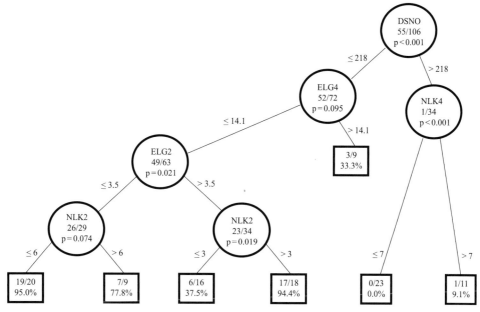

Figure 4.6. Classification and regression tree (CART) for Pacific coast Surf Scoters breeding in the northern boreal forest. The explanatory variable being split, the number of Surf Scoter breeding locations in the numerator and number of total locations (nesting areas and random locations) in the denominator, and the p-value associated with the split are included in each circle. Splitting values are indicated along the branches, and boxes indicate terminal node percentages of total locations comprised of Surf Scoter breeding locations. Splitting variables include distance to snowline (DSNO), elevation gradient (m/km) at 4 × 4 km (ELG4) and 2 × 2 km (ELG2), and number of lakes at 4 × 4 km (NLK4) and 2 × 2 km (NLK2).

individuals settled at nesting areas within a day of 30 May. Birds from different wintering areas were intermixed on the breeding grounds. A subsample of eight marked birds located in consecutive years showed strong nesting area fidelity.

Breeding Synchrony and Breeding Strategies

We found remarkable breeding synchrony for most birds regardless of their wintering origin and differences in distances from the breeding grounds. Birds wintering in Mexico were the only ones that arrived slightly later, although their sample size was small. Surf Scoters began arriving to their NBF breeding grounds around 24 May (JD = 144 ± 0.7: De La Cruz et al. 2009) and flew to nesting areas along the northeastern edge of their breeding range. They settled at nesting areas after only five days during the ice and snow breakup period, when availability of open water and food resources would be uncertain. This short pre-breeding period was similar to Barnacle Geese (*Branta leucopsis*), which had a pre-breeding period of <7 days (Prop et al. 2003).

Although early-arriving migrants may be more likely to breed successfully (Cooke et al. 1995, LePage et al. 2000), timing of migration must fit a narrow time window to provide maximal reproductive success (Drent et al. 2003). Surf Scoters must depart from coastal staging areas on a schedule to optimize their breeding opportunities, but they seem to have few proximate clues that would help predict interior weather conditions more than 1,000 km away. Significant capital investment is likely required for Surf Scoters to reproduce successfully in light of the unpredictable pre-breeding resources and the short Arctic summer period (Klaassen et al. 2006). On a capital–income breeder continuum (Klaassen et al. 2006), Surf Scoters are expected to be on the capital breeder side since they have a short breeding period, unpredictable food resources, and high predation risk (Jonsson 1997). Schmutz et al. (2006) examined isotope ratios and found that marine food from coastal stopovers was used for incubation by Emperor Geese (*Chen canagica*), but eggs had mixed ratios from exogenous and endogenous sources. Similarly, White-winged Scoters were reported to rely on exogenous resources for pre-laying and laying periods (Brown and Frederickson 1986).

In one of the few other studies to determine settling dates for waterfowl in the NBF, Ely et al. (2006) found earlier nest initiation (12 May) with less synchrony (8–16 May) for Tule Greater White-fronted Geese (*Anser albifrons elgasi*). However, this study was limited to VHF-marked individuals and defined nesting in upper Cook Inlet of central Alaska as those individuals located repeatedly for 1–2 weeks. Furthermore, Tule Greater White-fronted Geese may have less synchronous settling dates because they nest at lower latitudes, likely have more time to select nesting sites, and are larger, with lower risk of predation than Surf Scoters. Predation caused 76% of nest failures for another diving duck species that breeds in the NBF, the Lesser Scaup (*Aythya affinis*; Walker et al. 2005). The wide variation in annual recruitment rates seen in Lesser Scaup (Walker et al. 2005) is probably common for Surf Scoters, since their average annual nest success is very low (Savard et al. 1998).

In all likelihood, many more of the marked Surf Scoters were prospecting for nest sites early in the breeding season but their attempts to nest failed (S. W. De La Cruz, unpubl. data). Several VHF-marked females were located in the breeding area and appeared to be nesting, but subsequent trips to find their nests revealed that these birds had moved from their original positions. We suspect that their nests had been depredated, particularly since high nest depredation rates were noted by other researchers in the NWT (S. Slattery, Ducks Unlimited Canada, pers. comm.). In addition, small mammal populations were low and avian predators were markedly higher starting in 2005, when the majority of scoters were marked, than in previous years (S. Carrier, NWT Renewable Resources and Economic Development, pers. comm.). Waterfowl and their eggs may have provided an attractive alternative food source for predators (Ackerman 2002, Brook et al. 2005).

Wintering and Breeding Area Connectivity

Many species of migratory birds segregate by sex and age during the winter, often along latitudinal gradients (Ketterson and Nolan 1983), but few studies have linked breeding and wintering area subpopulations. Latitudinal variation of wintering Arctic nesting geese from different breeding grounds has been described (Ely and Takekawa 1996), and

variation in migration distances for Pacific Black Brant (*Branta bernicla nigricans*) migrating from wintering to breeding areas also has been documented (Schamber et al. 2007). Migration distance of Pacific Black Brant may be related to distribution of the eelgrass beds (*Zostera* spp.) that provide their primary food resource (Ward et al. 2005, Lindberg et al. 2007). Eelgrass beds also provide spawning habitat for Pacific herring (*Clupea pallasi*), and herring roe is a food resource consumed preferentially by Surf Scoters (Lewis et al. 2005).

We did not see a parallel relationship among latitudes at wintering and breeding areas where birds from more southerly areas were found at lower latitudes on the breeding grounds. We found that Surf Scoters from different wintering areas were randomly distributed on the breeding grounds. Surf Scoters from wintering areas were intermixed, although many PS birds migrated separately through the interior rather than following the coast (De La Cruz et al., 2009).

For the eight birds that were located nesting in consecutive years, we found philopatry and high nesting area fidelity. Female waterfowl show a high degree of nesting area fidelity, and the "local-knowledge" hypothesis suggests that females obtain an advantage in terms of food resources, distribution of conspecifics, and predator activity in an area (Rohwer 1992). For species such as sea ducks that form pairs in the winter, winter site fidelity may also be observed (Robertson and Cooke 1999). During our studies, few marked Surf Scoters moved among wintering areas during a single season, but one adult female marked in SF during 2005 flew to SQ during the winter of 2006. Complementary genetic analyses may be useful to further clarify the genetic structuring of these wintering populations.

Nesting Area Selection

Selection of breeding habitats by birds is considered to be hierarchical (Johnson 1980, Jones 2001), with different processes affecting different scales (Wiens 1989). We applied an unconstrained design that assumed selection varied at different scales (Battin and Lawler 2006) and focused our analyses at scales that we felt were appropriate for the accuracy of satellite-transmitter data. Compared with random locations in the NBF, both the logistic regression and CART analyses found consistent evidence that Surf Scoter nesting areas were located closer to the snowline and in areas with lower elevational gradients and more lakes (4 × 4 km scale). Both analyses also detected effects associated with lakes at the smaller scale (2 × 2 km); however, the CART suggests this effect may interact with elevation gradient rather than stand alone as a main effect. Our logistic regression also suggests that Surf Scoters selected locations farther from the treeline and at higher latitudes. The CART did not confirm these additional effects; however, after partitioning the data with respect to the strongest predictors, there may have been too little data within any of the partitions for the CART to identify further effects.

Sea ducks are the most northerly distributed ducks (Goudie et al. 1994), and we hypothesized that snow conditions could be critical in determining nesting area. In fact, distance to snowline was the greatest single factor in the AIC score. The date when an area was snow free and the variation around that date were only ranked highly in backward selection, possibly because these variables were not completely independent and birds were selecting open habitats at finer scales. Because elevation was not selected and elevation gradient was highly ranked at both scales, Surf Scoters were likely selecting areas where topographic variation resulted in earlier openings in snow cover in some areas. Habitat selection may vary when available habitats vary temporally (Warnock and Takekawa 1995, Dailey et al. 2007), and we were able to account for this changing availability of nesting areas through use of daily ice and snow cover data.

Breeding propensity, or the likelihood that birds will find adequate habitat available to initiate a breeding attempt (Petrie et al. 2000), is probably highly variable for Surf Scoters because the spring thaw varies each year. It may be especially variable for those species that nest in unpredictable and highly heterogeneous environments such as the Arctic. For example, a large proportion of arctic nesting geese may fail to breed in years of late snow melt (Barry 1962, Prop and Devries 1993, Ganter and Boyd 2000). Delayed snow melt can prevent access to nesting sites and impair acquisition of nutrients for egg formation of Lesser Snow Geese (*Chen c. caerulescens*; Ganter and Cooke 1996). Breeding propensity in Greater Snow Geese (*C. c. atlantica*) varied from 0.17 to 1.00, and spring snow cover was a critical determinant (Reed et al. 2004). Most females (>80%) bred when snow cover was low, but few (<30%) bred when snow cover was extensive.

Our CART analysis suggested that nesting areas were found in areas where there were 3–6 lakes at the 2 × 2 km scale. At lower elevation gradients (≤3.5 m/km), similar proportions of nesting areas were found regardless of number of lakes, possibly because the area was more homogeneous, but at higher elevation gradients (>3.5 m/km), more nests were located when there were more lakes that could support food resources. Similarly, Perry et al. (2006) found that Surf Scoters from the Atlantic coast were located in areas with large wetlands (22 ha) associated with small rivers and when there were 22 lakes found within 1 km of the presumed breeding lake. Surf Scoters are reported to use lakes less than 10 ha in size that are shallow (Decarie et al. 1995), possibly to avoid competition with fish that consume invertebrates or to avoid larger fish that may consume young ducklings (Mallory et al. 1994). Unfortunately, we were unable to assess water depth from the satellite data to examine if shallow depths were selected.

A few other variables were highlighted by the AIC analysis. Nesting area selection was positively related to latitude, supporting our finding that Surf Scoters selected nesting areas in the northeast part of the breeding range. Surf Scoters migrated past many apparently suitable areas to the southwest of the snowline, but the reason why they avoid those areas is not clear. Perrin's hypothesis (Drent 2006) suggests that timing of nesting activities is a compromise between survival of the adult and food resource availability for the young. Nesting area selection may be driven by security and food availability for Surf Scoters, since this long-lived species seems to have low annual nest success (Newton 1989, Clutton-Brock 1998, Savard et al. 1998). Ducklings hatch nearly a month after breakup, and they may benefit from greater availability of macroinvertebrates (Oswood et al. 2006). Nest predators may be less abundant in areas farther north. White-winged Scoters (*Melanitta fusca*) do nest in areas farther south, but they are larger, which may deter some predators, and often use islands to avoid other predators (Traylor et al. 2004, Perry et al. 2006).

The AIC analysis also suggested that nesting areas were negatively related to treeline and positively related to tree cover. Treelines are sensitive to changes in climate, as well as being proxies for biotic changes (reviewed in Payette 2007). Surf Scoters did not settle near treeline, possibly because tree and ground cover was too sparse to provide adequate cover against predators or severe weather events. Perry et al. (2006) also found that most Surf Scoters from Atlantic coast wintering areas bred in forested areas. Surf Scoters may select smaller areas with adequate cover rather than larger, contiguous forested blocks. Fire frequency is high in the NBF (Kasischke et al. 2006), and the availability of contiguous forest blocks may be highly variable.

Implications of Climate Change for Surf Scoters in the Boreal Forest

The synchrony of breeding and the relation of nesting locations to snowline suggest that reproductive timing may be constrained by resource conditions. If adults are nesting earlier to maximize their own food resources, it may be that their breeding productivity depends on resource availability for their ducklings. Climate change is most pronounced at high latitudes (Serreze et al. 2000), and global warming may result in drastic changes in the timing of invertebrate blooms in the boreal forest. At Delta Marsh, Manitoba, waterfowl had the highest proportion of earlier arrivals among migratory birds, and arrival date was correlated with temperature (Murphy-Klassen et al. 2005). Weather at stopover areas and at the final destination influences timing of bird migration (Richardson 1978). Dates of ice break-up are good predictors because open water plays a primary role in migration of waterfowl (Murphy-Klassen et al. 2005).

The spring mismatch hypothesis (Visser et al. 1998, Drever and Clark 2007) suggests that global warming may result in earlier or protracted blooms of invertebrates, resulting in less food available for ducklings during critical periods of early growth. Haszard (2004) found Surf Scoters pairs and broods in wetlands with more abundant food, but their use of wetlands was not clearly related to wetland amphipods. However, wetlands used by Lesser Scaup in the same region were strongly related to abundance of amphipods (Fast et al. 2004). Surf Scoters are relatively unique in that many ducklings gather together in large creches and largely fend for themselves within a few days after leaving the nest (Savard et al. 1998). Thus, the availability of abundant invertebrate prey in brood-rearing areas may be critical for duckling survival.

Global warming has been cited as a conservation concern for bird species throughout the world

(McCarty 2001) and may have greatest impacts on the most northerly nesting species, such as sea ducks. Birds respond by expanding their range northward, advancing breeding dates, or varying their timing of breeding to match phenology of their prey (Ward 1992, Drever and Clark 2007). Individuals or populations that are not able to adjust to warming may suffer reduced productivity because of limitations in egg production, or because of conflicts in wintering ground or migration cues compared with breeding timing (Both and Visser 2001). Under the individual optimization hypothesis (Drent 2006), timing of breeding is controlled by environmental factors, but endogenous controls (Gwinner 1996) may limit the ability of Surf Scoters to adapt to climatic changes.

Conservation Threats

A better understanding of scoter breeding ecology will aid managers in making informed land use decisions as development pressure increases. For example, our results identify a much smaller area where effort could be focused to examine declining populations of Surf Scoters from the Pacific coast. The need for this information is underscored by the imminent oil, gas, and diamond mining development planned in the NWT (Government of the Northwest Territories; Industry, Tourism, and Investment, http://www.iti.gov.nt.ca/index.html). Loss of habitat from logging, mining, and hydroelectric power production has been suspected of affecting Surf Scoters from the Atlantic coast (Perry et al. 2006).

Sustainable harvest rates in sea ducks may be lower than in many other species (Goudie et al. 1994), since sea ducks tend to be K-selected (Eadie et al. 1988). Sea ducks have deferred sexual maturity, low annual recruitment to breeding age, variable rates of non-breeding by adults and high annual adult survival (Goudie et al. 1994). Infrequent Arctic ice events may cause mass mortality (Barry 1968) or affect body condition and fitness of birds (Goudie and Ankney 1986). Thus, subtle changes in the frequency of catastrophic events may greatly reduce population levels over time.

ACKNOWLEDGMENTS

This study was funded by the U.S. Geological Survey (USGS) Western Ecological Research Center under the Coastal Ecosystems and Land-Sea Interface

Program. Additional assistance was provided by the CALFED Ecosystem Restoration Program Mercury Project, the Dennis Raveling Chair at the University of California, and the Sea Duck Joint Venture and NASA Signals of Spring. T. Bowman (SDJV) encouraged us to integrate our data, and we acknowledge the support of D. Mulcahy, S. Iverson, K. Brodhead, P. Fontaine, E. Bohman, C. Eldermire, A. Keech, S. Duarte-Etchart, S. Duarte, E. Lok, D. Rizzolo, K. Sage, and M. Shepherd (SQ); D. Gaube, L. Terrazas, J. Anhalt, C. Kereki, J. Chastant, H. Goyert, J. Wasley, J. Seyfried, P. Gibbons, M. Nagendran, C. Scott, P. Tucker, C. Salido (SF); Tom Cyra, Joe Gaydos, Briggs Hall, Dyanna Lambourn, Don Kraege, Bryan Murphie, and Greg Schirato (PS); the Canadian Wildlife Service of Environment Canada and National Science and Engineering Research Council (NSERC) strategic grant STPGP246079-01, the Centre for Wildlife Ecology, Simon Fraser University, and field assistance of E. Anderson, B. Bartzen, T. Bowman, S. Coulter, R. Dickson, G. Grigg, S. Iverson, D. Lacroix, T. Lewis, R. Lis, and R. Zydelis, D. Mulcahy, and M. McAdie for radio implantation (SG). On the breeding grounds, we were grateful for the assistance of C. Babcock and A. Fowler (UC Davis), who led the initial surveys in 2003, and K. Farke and I. Woo (USGS), who assisted in 2005. Also, special thanks to R. King, FWS Migratory Bird Management (retired) for flight support, J. Hines (CWS, Yellowknife, NWT) for assistance with housing and information, D. Douglas (USGS Alaska Science Center) for advice on remote-sensing data sources, and E. Butterworth (Ducks Unlimited Canada) for the western boreal forest and treeline coverages. We thank M. Mueller, S. Iverson, M. Casazza, and T. Bowman for helpful comments on earlier drafts. This work was conducted following guidance from Animal Care and Use Committees and with permits from California Department of Fish and Game, Canadian Wildlife Service, Northwest Territories RWED, U.S. Fish and Wildlife Service, USGS Bird Banding Laboratory, and Washington Department of Fish and Wildlife. We also thank the numerous local NWT communities that provided helpful information and permitted access to remote areas. Use of trade names does not imply government endorsement.

LITERATURE CITED

Ackerman, J. T. 2002. Of mice and mallards: positive indirect effects of coexisting prey on waterfowl nest success. Oikos 98:469–480.

Anderson, D. R., K. P. Burnham, and W. L. Thompson. 2000. Null hypothesis testing: problems, prevalence, and an alternative. Journal of Wildlife Management 64:912–923.

Barry, T. W. 1962. Effects of late seasons on Atlantic Brant reproduction. Journal of Wildlife Management 26:19–26.

Barry, T. W. 1968. Observations on natural mortality and native use of eider ducks along the Beaufort Sea coast. Canadian Field-Naturalist 82:140–144.

Battin, J., and J. J. Lawler. 2006. Cross-scale correlation and the design and analysis of avian habitat selection studies. Condor 108:59–70.

Beyer, H. L. 2004. Hawth's Analysis Tools for ArcGIS. <http://www.spatialecology.com/htools>.

Both, C., and M. E. Visser. 2001. Adjustment to climate change is constrained by arrival date in a long-distance migrant bird. Nature 411:296–298.

Brieman, L., J. H. Friedman, R. A. Olshen, and C. J. Stone. 1984. Classification and regression trees. Wadsworth and Brooks, Monterey, CA.

Brook, R. W., D. C. Duncan, J. E. Hines, S. Carriere, and R. G. Clark. 2005. Effects of small mammals cycles on productivity of boreal ducks. Wildlife Biology 11:3–11.

Brown, P. W., and L. H. Frederickson. 1986. Food-habits of breeding White-winged Scoters. Canadian Journal of Zoology 64:1652–1654.

Burnham, K. P., and D. R. Anderson. 2002. Model selection and multimodel inference: a practical information-theoretic approach. 2nd ed. Springer Science+Business Media, Inc., New York, NY.

Carlton, J. T., J. K. Thompson, L. E. Schemel, and F. H. Nichols. 1990. Remarkable invasion of San Francisco Bay (California, USA) by the Asian clam Potamocorbula amurensis. I: Introduction and dispersal. Marine Ecology Progress Series 66:81–94.

Chapin, F. S., III, J. Yarie, K. Van Cleve, and L. A. Viereck. 2006. The conceptual basis of LTER studies in the Alaskan boreal forest. Pp. 3–11 in F. S. Chapin III, M. W. Oswood, K. Van Cleve, L. A. Viereck, and D. L. Verbyla (editors), Alaska's changing boreal forest. Oxford University Press, New York, NY.

Clutton-Brock, T. H. 1988. Reproductive success. University of Chicago Press, Chicago, IL.

Cohen, A. N., and J. T. Carlton. 1995. Nonindigenous aquatic species in a United States estuary: a case study of the biological invasions of the San Francisco Bay and Delta. Report to U.S. Fish and Wildlife Service and National Sea Grant College Program, Connecticut Sea Grant.

Cooke, F., R. F. Rockwell, and D. B. Blank. 1995. The Snow Geese of La Perouse Bay: natural selection in the wild. Oxford University Press, Oxford, UK.

Cuzick, J., and R. Edwards. 1990. Spatial clustering for inhomogeneous populations. Journal of the Royal Statistical Society B. 52:73–104.

Dailey, M., A. I. Gitelman, F. L. Ramsey, and S. Starcevich. 2007. Habitat selection models to account for seasonal persistence in radio telemetry data. Environmental and Ecological Statistics 14:55–68.

Day, R., J. H. Ware, D. Wartenberg, and M. Zelen. 1989. An investigation of a reported cancer cluster in Randolf, Massachusetts. Journal of Clinical Epidemiology 42:137–150.

De La Cruz, S. E. W., J. Y. Takekawa, M. T. Wilson, D. R. Nysewander, J. R. Evenson, D. Esler, W. S. Boyd, and D. H. Ward. 2009. Spring migration routes and chronology of Surf Scoters (*Melanitta perspicillata*): a synthesis of Pacific coast studies. Canadian Journal of Zoology 87:1069–1086.

De'ath, G., and K. E. Fabricius. 2000. Classification and regression trees: a powerful yet simple technique for ecological data analysis. Ecology 81:3178–3192.

Decarie, R., F. Morneau, D. Lambert, S. Carriere, and J.-P. L. Savard. 1995. Habitat use of brood-rearing waterfowl in subarctic Quebec. Arctic 48:383–390.

Devries, J. H., J. J. Citta, M. S. Lindberg, D. W. Howerter, and M. G. Anderson. 2003. Breeding-season survival of mallard females in the Prairie Pothole Region of Canada. Journal of Wildlife Management 67:551–563.

Drent, R. 2006. The timing of birds' breeding seasons: the Perrins hypothesis revisited especially for migrants. Ardea 94:305–322.

Drent, R., C. Both, M. Green, J. Madsen, and T. Piersma. 2003. Pay-offs and penalties of competing migratory schedules. Oikos 103:274–292.

Drever, M. C., and R. G. Clark. 2007. Spring temperature, clutch initiation date and duck nest success: a test of the mismatch hypothesis. Journal of Animal Ecology 76:139–148.

Eadie, J. M., F. P. Kehoe, and T. D. Nudds. 1988. Pre-hatch and post-hatch brood amalgamation in North American Anatidae: a review of hypotheses. Canadian Journal of Zoology 66:1709–1721.

Ely, C. R., K. S. Bollinger, R. V. Densmore, T. C. Rothe, M. J. Petrula, J. Y. Takekawa, and D. L. Orthmeyer. 2007. Reproductive strategies of northern geese: why wait? Auk 124:594–605.

Ely, C. R., K. S. Bollinger, J. W. Hupp, D. V. Derksen, J. Terenzi, J. Y. Takekawa, D. L. Orthmeyer, T. C. Rothe, M. J. Petrula, and D. R. Yparraguirre. 2006. Traversing a boreal forest landscape: summer movements of Tule Greater White-fronted Geese. Waterbirds 29:43–55.

Ely, C. R., and J. Y. Takekawa. 1996. Geographic variation in migratory behavior of Greater White-fronted Geese (*Anser albifrons*). Auk 113:889–901.

ESRI. 1996. ArcView GIS: using ArcView GIS. Environmental Systems Research Institute, Inc., Redlands, CA.

Fast, P. L. F., R. G. Clark, R. W. Brook, and J. E. Hines. 2004. Patterns of wetland use by brood-rearing Lesser Scaup in northern boreal forest of Canada. Waterbirds 27:177–182.

Ganter, B., and H. Boyd. 2000. A tropical volcano, high predation pressure and the breeding biology of arctic waterbirds: a circumpolar review of breeding failure in the summer of 1992. Arctic 53:289–305.

Ganter, B., and F. Cooke. 1996. Pre-incubation feeding activities and energy budgets of Snow Geese: can food on the breeding grounds influence fecundity? Oecologia 106:153–165.

Goudie, R. I., and C. D. Ankney. 1986. Body size, activity budgets, and diets of sea ducks wintering in Newfoundland. Ecology 67:1475–1482.

Goudie, R. I., A. V. Kondratyev, S. Brault, M. R. Petersen, B. Conant, and K. Vermeer. 1994. The status of sea ducks in the north Pacific Rim: toward their conservation and management. Transactions of the 59th North American Wildlife and Natural Resources Conference 59:27–49.

Gwinner, E. 1996. Circannual clocks in avian reproduction and migration. Ibis 138:47–63.

Hansen, M., R. DeFries, J. R. Townshend, M. Carroll, C. Dimiceli, and R. Sohlberg. 2001. Vegetation continuous fields MOD44B, 2001 percent tree cover. Collection 4. University of Maryland. College Park, MD.

Haszard, S. 2004. Habitat use by White-winged Scoters (*Melanitta fusca*) and Surf Scoters (*Melanitta perspicillata*) in the Mackenzie Delta region, Northwest Territories. Master's thesis, University of Saskatchewan, Saskatoon, SK.

Hodges, J. I., J. G. King, B. Conant, and H. A. Hanson. 1996. Aerial surveys of waterbirds in Alaska 1957–94: population trends and observer variability. Information and Technology Report 4. U.S. National Biological Service and U.S. Fish and Wildlife Service.

Hothorn, T., K. Hornik, and A. Zeileis. 2006. Unbiased recursive partitioning: a conditional inference framework. Journal of Computational and Graphical Statistics. 15:651–674.

Hothorn, T., F. Leisch, and A. Zeileis. 2008. Model-tools: tools and classes for statistical models. R package version 0.2-15.

Iverson, S. A., W. S. Boyd, D. Esler, D. M. Mulcahy, and T. D. Bowman. 2006. Comparison of the effects and performance of four types of radiotransmitters for use with scoters. Wildlife Society Bulletin 34:656–663.

Jessen, R. L. 1981. Special problems with diving ducks. Proceedings of the International Waterfowl Symposium 4:139–148.

Johnson, D. H. 1980. The comparison of usage and availability measurements for evaluating resource preference. Ecology 61:65–71.

Jones, J. 2001. Habitat selection studies in avian ecology: a critical review. Auk 118:557–562.

Jonsson, J. 1997. Habitat selection studies in avian ecology: a critical review. Auk 118:557–562.

Kaiser, G. W., A. E. Derocher, S. Crawford, M. J. Gill, and I. A. Manley. 1995. A capture technique for Marbled Murrelets in coastal inlets. Journal of Field Ornithology 66:321–333.

Kasischke, E. S., T. S. Rupp, and D. L. Verbyla. 2006. Fire trends in the Alaskan boreal forest. Pp. 285–301 in F. S. Chapin III, M. W. Oswood, K. Van Cleve, L. A. Viereck, and D. L. Verbyla (editors), Alaska's changing boreal forest. Oxford University Press, New York, NY.

Ketterson, E. D., and V. Nolan, Jr. 1983. The evolution of differential bird migration. Current Ornithology 1:357–402.

Klaassen, M., K. F. Abraham, R. L. Jeffries, and M. Vrtiska. 2006. Factors affecting the site of investment, and the reliance on savings for Arctic breeders: the capital-income dichotomy revisited. Ardea 94:371–384.

Korschgen, C. E., K. P. Kenow, A. Gendron-Fitzpatrick, W. L. Green, and F. J. Dein. 1996. Implanting intra-abdominal radio transmitters with external whip antennas in ducks. Journal of Wildlife Management 60:132–137.

LePage, D., G. Gauthier, and S. Menu. 2000. Reproductive consequences of egg-laying decisions in Snow Geese. Journal of Animal Ecology 69:414–427.

Lewis, T. L., D. Esler, W. S. Boyd, and R. Zydelis. 2005. The nocturnal foraging behaviors of wintering Surf Scoters and White-winged Scoters. Condor 107:636–646.

Lindberg, M. S., D. H. Ward, T. L. Tibbetts, and J. Roser. 2007. Winter movement dynamics of black brant, Journal of Wildlife Management 71:534–540.

Linville, R. G., S. N. Luoma, L. Cutter, and G. Cutter. 2002. Increased selenium threat as a result of the invasion of the exotic bivalve Potamacorbula amurensis in the San Francisco Bay-Delta. Aquatic Toxicology 57:51–64.

Maisonneuve, C., L. Belanger, D. Bordage, B. Jobin, M. Grenier, J. Beaulieu, S. Gabor, and B. Filion. 2006. American Black Duck and Mallard breeding distribution and habitat relationships along a forest-agriculture gradient in southern Quebec. Journal of Wildlife Management 70:450–459.

Mallory, M. L., P. J. Blancher, P. J. Weatherhead, and D. K. McNichol. 1994. Presence or absence of fish as a cue to macroinvertebrate abundance in boreal wetlands. Hydrobiologia 279/280:345–351.

Manly, B. F. J., L. L. McDonald, D. L. Thomas, T. L. McDonald, and W. P. Erickson. 2002. Resource selection by animals: statistical design and analysis for field studies. 2nd ed. Kluwer Academic Publishers, Dordrecht, Netherlands.

McCarty, J. P. 2001. Ecological consequences of recent climate change. Conservation Biology 15:320–331.

Miller, M. R., J. Y. Takekawa, J. P. Fleskes, D. L. Orthmeyer, M. L. Casazza, and W. M. Perry. 2005. Spring migration of northern pintails from California's Central Valley wintering area tracked with satellite telemetry: routes, timing, and destinations. Canadian Journal of Zoology 83:1314–1332.

Murphy-Klassen, H. M., T. J. Underwood, S. G. Sealy, and A. A. Czyrnyj. 2005. Long-term trends in spring arrival dates of migrant birds at Delta Marsh, Manitoba in relation to climate change. Auk 122:1130–1148.

Newton, I. 1989. Lifetime reproduction in birds. Academic Press, London, UK.

Nichols, F. H., J. E. Cloern, S. N. Luoma, and D. H. Peterson. 1986. The modification of an estuary. Science 231:567–573.

NOAA/NESDIS/OSDPD/SSD. 2006. IMS daily northern hemisphere snow and ice analysis at 4 km and 24 km resolution. National Snow and Ice Data Center, Boulder, CO.

Nysewander, D. R., J. R. Evenson, B. L. Murphie, and T. A. Cyra. 2004. Trends observed for selected marine bird species during 1993–2002 winter aerial surveys. In T. Droscher and D. A. Fraser (editors), Proceedings of the Georgia Basin/Puget Sound Research Conference. <http://www.psat.wa.gov/Publications/03_proceedings/start.htm>.

Oswood, M. W., N. F. Hughes, and A. M. Milner. 2006. Running waters of the Alaskan boreal forest. Pp. 147–167 in F. S. Chapin III, M. W. Oswood, K. Van Cleve, L. A. Viereck, and D. L. Verbyla (editors), Alaska's changing boreal forest. Oxford University Press, New York, NY.

Payette, S. 2007. Contrasted dynamics of northern Labrador tree lines caused by climate change and migrational lag. Ecology 88:770–780.

Perry, M. C., D. M. Kidwell, A. M. Wells, E. J. R. Lohnes, P. C. Osenton, and S. H. Altmann. 2006. Characterization of breeding habitats for Black and Surf Scoters in the eastern boreal forest and subarctic regions of Canada. Pp. 80–89 in A. Hansen, J. Kerekes, and J Paquet. Limnology and waterbirds 2003. Canadian Wildlife Service Technical Report Series No. 474.

Petrie, M. J., R. D. Drobney, and D. T. Sears. 2000. Mallard and Black Duck breeding parameters in New Brunswick: a test of the reproductive rate hypothesis. Journal of Wildlife Management 64: 832–838.

Prop, J., J. M. Black, and P. Shimmings. 2003. Travel schedules to the high Arctic: Barnacle Geese trade-off the timing of migration with accumulation of fat deposits. Oikos 103:403–414.

Prop, J., and J. Devries. 1993. Impact of snow and food conditions on the reproductive performance of barnacle geese *Branta leucopsis*. Ornis Scandinavica 24:110–121.

R Development Core Team. 2008. R: a language and environment for statistical computing. R Foundation for Statistical Computing, Vienna, Austria. <http://www.R-project.org>.

Reed, E. T., G. Gauthier, and J.-F. Giroux. 2004. Effects of spring conditions on breeding propensity of Greater Snow Goose females. Animal Biodiversity and Conservation 27:35–46.

Rempel, R. S., K. F. Abraham, T. R. Gadawski, S. Gabor, and R. K. Ross. 1997. A simple wetland habitat classification for boreal forest waterfowl. Journal of Wildlife Management 61:746–757.

Richardson, W. J. 1978. Timing and amount of bird migration in relation to weather: a review. Oikos 30:224–272.

Robertson, G. J., and F. Cooke. 1999. Winter philopatry in migratory waterfowl. Auk 116:20–34.

Rohwer, F. C. 1992. The evolution of reproductive patterns in waterfowl. Pp. 486–539 *in* B. D. J. Batt, A. D. Afton, M. G. Anderson, C. D. Ankney, D. H. Johnson, J. A. Kadlec, and G. L. Krapu (editors), Ecology and management of breeding waterfowl. University of Minnesota Press. Minneapolis, MN.

Rosing, M. N., M. Ben-David, and R. P. Perry. 1998. Analysis of stable isotope data: a K nearest-neighbors randomization test. Journal of Wildlife Management 62:380–388.

SAS Institute. 2004. SAS OnlineDoc® 9.1.2. SAS Institute, Inc., Cary, NC.

Savard, J.-P. L., and P. Lamothe. 1991. Distribution, abundance, and aspects of breeding ecology of Black Scoters, *Melanitta nigra*, and Surf Scoters, *M. perspicillata*, in northern Quebec. Canadian Field-Naturalist 105:488–496.

Savard, J.-P. L., D. Bordage, and A. Reed. 1998. Surf Scoter (*Melanitta perspicillata*). A. Poole and F. Gill (editors), The birds of North American No. 363. The Birds of North America, Inc., Philadelphia, PA.

Schamber, J. L., J. S. Sedinger, D. H. Ward, and K. R. Hagemeier. 2007. Latitudinal variation in population structure of wintering Pacific Black Brant. Journal of Field Ornithology 78:74–82.

Schilling, M. F. 1986. Multivariate two-sample tests based on nearest neighbors. Journal of the American Statistical Association 81:799–806.

Schmutz, J. A., K. A. Hobson, and J. A. Morse. 2006. An isotopic assessment of protein from diet and endogenous stores: effects on egg production and incubation behavior of geese. Ardea 94:385–397.

Sea Duck Joint Venture (SDJV) Management Board. 2001. Sea Duck Joint Venture strategic plan: 2001–2006. SDJV Continental Team. Unpublished report. USFWS, Anchorage, AK; CWS, Sackville, NB.

Serreze, M. C., J. E. Walsh, F. S. Chapin III, T. Osterkamp, M. Dyurgerov, V. Romanovsky, W. C. Oechel, J. Morison, T. Zhang, and R. G. Barry. 2000. Observational evidence of recent change in the northern high-latitude environment. Climatic Change 46:159–207.

Shugart, H. H., R. Leemans, and G. B. Bonan (editors). 1972. A systems analysis of the global boreal forest. Cambridge University Press, Cambridge, UK.

Smith, G. W. 1995. A critical review of the aerial and ground surveys of breeding waterfowl in North America. Biological Science Report 5. National Biological Service, Washington, DC.

Smol, J. P., and M. S. V. Douglas. 2007. Crossing the final ecological threshold in high Arctic ponds. Proceedings of the National Academy of Sciences 104:12395–12397.

Soja, A. J., N. M. Tchebakova, N. H. F. French, M. D. Flannigan, H. H. Shugart, B. J. Stocks, A. I. Sukhinin, E. I. Parfenova, F. S. Chapin III, and P. W. Stackhouse, Jr. 2007. Climate-induced boreal forest change: predictions versus current observations. Global and Planetary Change 56:274–296.

Steidl, R. J. 2006. Model selection, hypothesis testing, and risks of condemning analytical tools. Journal of Wildlife Management 70:1497–1498.

Stephens, P. A., S. W. Buskirk, G. D. Hayward, and C. Martinez del Rio. 2005. Information theory and hypothesis testing: a call for pluralism. Journal of Applied Ecology 42:4–12.

Traylor, J. J., R. T. Alisauskas, and F. P. Kehoe. 2004. Nesting ecology of White-winged Scoters (*Melanitta fusca deglandi*) at Redberry Lake, Saskatchewan. Auk 121:950–962.

U.S. Fish and Wildlife Service (USFWS). 2002. Waterfowl population status, 2002. U.S. Department of the Interior, Washington, DC.

Visser, M. E., A. J. Van Noordwijk, J. M. Tinbergen, and C. M. Lessels. 1998. Warmer springs lead to mistimed reproduction in Great Tits (*Parus major*). Proceedings of the Royal Society of London Series B, 265:1867–1870.

Walker, J., M. S. Lindberg, M. C. MacCluskie, M. J. Petrula, and J. S Sedinger. 2005. Nest survival of scaup and other ducks in the boreal forest of Alaska. Journal of Wildlife Management 69:582–591.

Ward, D. H., A. Reed, J. S. Sedinger, J. M. Black, D. V. Derksen, and P. M. Castelli. 2005. North American Brant: effect of changes in habitat and climate on population dynamics. Global Change Biology 11:869–880.

Ward, J. V. 1992. Aquatic insect ecology, Vol. 1: Biology and habitat. John Wiley and Sons, Inc., New York, NY.

Warnock, S. E., and J. Y. Takekawa. 1995. Habitat preferences of wintering shorebirds in a temporally changing environment: Western Sandpipers in the San Francisco Bay estuary. Auk 112:920–930.

Webster, M. S., P. P. Marra, S. M. Haig, S. Bencsch, and R. T. Holmes. 2002. Links between worlds: unraveling migratory connectivity. Trends in Ecology and Evolution 17:76–83.

White , G. C., and R. A. Garrott. 1990. Analysis of wildlife radio-tracking data. Academic Press, San Diego, CA.

Wiens, J. A. 1989. Spatial scaling in ecology. Functional Ecology 3:385–397.

CHAPTER FIVE

Geospatial Modeling of Abundance with Breeding Bird Atlas Data

Andrew R. Couturier

Abstract. Recent studies demonstrate that North America's boreal forest region provides critical nesting habitat for countless songbirds, shorebirds, and waterfowl. Despite its obvious importance to the avifauna of the Americas, specific detail concerning the distribution and abundance of birds in the boreal forest region is largely lacking. The boreal forest is changing rapidly, as development encroaches further north in the form of logging, mining, oil and gas exploration, and human settlement. The cumulative impacts of these and other activities, against the backdrop of a changing climate, underscore the need to quantify the distribution and abundance of birds at scales that are detailed enough to inform landscape planning and management decisions. Data on relative bird abundance collected for the Ontario Breeding Bird Atlas can help fill this information need. This paper reports on the potential value of the relative abundance information for conservation planning in the boreal region of the province.

Key Words: abundance, birds, boreal forest, breeding bird atlas, conservation applications, kriging, Ontario, point counts.

The boreal forest region is a globally important forest ecosystem that provides a wide variety of services and functions, including biogeochemical and hydrological services, important wildlife habitat, lands for First Nations peoples, and many others (Canadian Boreal Initiative 2005). The boreal forest spans large expanses of northern countries including Canada, Russia, Finland, Sweden, the United States (Alaska), and others. The boreal forest encompasses nearly 50% of Canada's land mass and, for the most part, remains sparsely populated and relatively intact. However, the region is under considerable development pressure from forestry, mining, and oil and gas extraction, underscoring the need for spatially explicit conservation frameworks.

Recent studies have demonstrated the importance of the North American boreal forest region to a wide variety of bird species (Blancher and Wells 2005; Wells and Blancher, this volume, chapter 2). At least 325 species are known to be dependent on the region during breeding, migration, or wintering seasons (Wells and Blancher, this volume, chapter 2).

Couturier, A. R. 2011. Geospatial modeling of abundance with breeding bird atlas data. Pp. 65–72 *in* J. V. Wells (editor). Boreal birds of North America: a hemispheric view of their conservation links and significance. Studies in Avian Biology (no. 41), University of California Press, Berkeley, CA.

Despite the importance of the boreal forest to birds, its remote and inaccessible nature has slowed progress in documenting the precise distribution and abundance of these species across the landscape, information that is essential for effective land use planning and decision making. This paper describes the value of atlas data for documenting patterns of distribution and abundance and also explores its potential to inform conservation planning in the boreal region of Ontario, Canada, using information from the recently published *Atlas of the Breeding Birds of Ontario* (Cadman et al. 2007).

DATA SOURCES AND METHODS

In North America, breeding bird atlas projects have traditionally mapped the distribution of breeding bird species across a region based on standardized observations of breeding activity within a predefined system of grid cells covering the region of interest, usually a province or state. Atlases typically run for five years, involve the efforts of thousands of volunteer scientists, are spearheaded by various configurations of not-for-profit organizations and governments, and are repeated on a 20-year cycle. Breeding bird atlas projects started in the U.K. in the 1960s and quickly became a central feature of biodiversity monitoring. Efforts in North America evolved gradually, with the first atlases appearing in the United States and Canada in the mid 1970s and early 1980s. Today, hundreds of bird atlas projects have been completed around the world, and in many regions a series of atlases have been completed that allow documentation of distributional changes through time (see Dunn and Weston 2008 for examples). In Canada, second breeding bird atlases have been published, or are in progress, for Alberta, Ontario, Quebec, and the Maritimes. In addition, first atlases are in progress for British Columbia and Manitoba. Plans for atlas projects in the remaining provinces (and possibly territories) are under consideration. In Canada, as in the U.K. 20 years ago, atlases are increasingly becoming key tools for monitoring and conservation.

At a minimum, breeding bird atlas projects compile information concerning the likelihood that a given species is breeding within a designated spatial unit. This "presence" information is very valuable for mapping the broad range of a species. A main limitation is that its relative abundance within that range is unknown: a sampling unit could contain one pair of a given species or 1,000 pairs, but would be recorded only as "present" in both instances. While abundance sampling is a long-established component of atlases in the U.K. and other European jurisdictions, it has only recently begun to be incorporated into atlas projects in North America.

Canada's first provincial breeding bird atlas project was completed in Ontario in 1985 and published in 1987 (Cadman et al. 1987). A second Ontario atlas was completed in 2005 and published in 2007 (Cadman et al. 2007). As with all repeat atlases, a primary objective was to duplicate coverage from the previous atlas so that changes in distribution could be assessed. Additionally, the second Ontario atlas was the first in North America to collect comprehensive, quantitative bird data in order to develop maps that depict relative abundance across broad geographic areas.

For abundance sampling, the Ontario atlas devised a system that consisted of randomly located roadside point counts, supplemented with a suitable number of off-road points in target habitats that were underrepresented along roadsides. More than 60,000 point counts were conducted during the five-year project, with near-complete coverage in the southern part of the province and good coverage in the road-accessible southern boreal forest. Coverage in the remote north, beyond the road system, was sparse and somewhat opportunistic, but adequate for abundance mapping (Fig. 5.1). Additional detail concerning the sampling methodology, coverage achieved, and results can be found in Cadman et al. (2007). All new atlas projects in Canada are now using the approach developed for Ontario, with appropriate regional adaptations. Increasingly, atlas projects in the United States are also moving to include quantitative sampling of some sort.

The Ontario Atlas used a spatial interpolation method—ordinary kriging—to convert discrete point count data values to continuous surfaces of relative abundance for 130 species (see Cadman et al. 2007 for methodological details). This work has led to new and exciting information for a large number of species in the province, and in particular for a variety of species that breed in the boreal

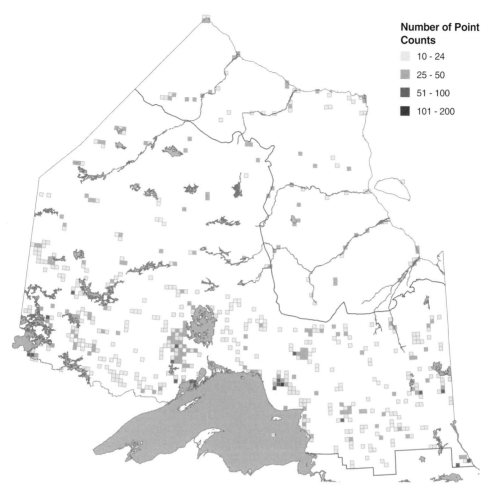

Figure 5.1. Number of point counts conducted per 10-km atlas square in northern Ontario. The map includes only those squares with at least 10 point counts, the minimum number deemed acceptable for inclusion in analyses and mapping. Source: Cadman et al. (2007).

Number of Point Counts
- 10 - 24
- 25 - 50
- 51 - 100
- 101 - 200

forest region of the province (see examples in Fig. 5.2). These maps demonstrate the power and value of relative abundance data. For example, the widespread and common Dark-eyed Junco (*Junco hyemalis*), previously assumed to be evenly distributed throughout its range, is actually far more abundant in the far north than in the southern boreal forest. Cadman et al. (2007) is replete with similar examples. The maps provide the most comprehensive—and often the first—look in the province's history at the relative abundance of many birds and thus fill an important information gap.

Relative abundance data have tremendous potential to inform conservation planning in the boreal forest region and elsewhere. From answering fundamental questions such as "where and how many" to complex spatial modeling and land allocation algorithms, atlas data have an important role to play. Some of the potential applications of atlas data, and quantitative data in particular, are elaborated below. See Pomerey et al. (2008) for a broader overview of atlas projects and their use in conservation.

DISCUSSION

A primary motivation for undertaking quantitative sampling of bird populations is to gain a better understanding of precisely where species are most abundant and where they are peripheral.

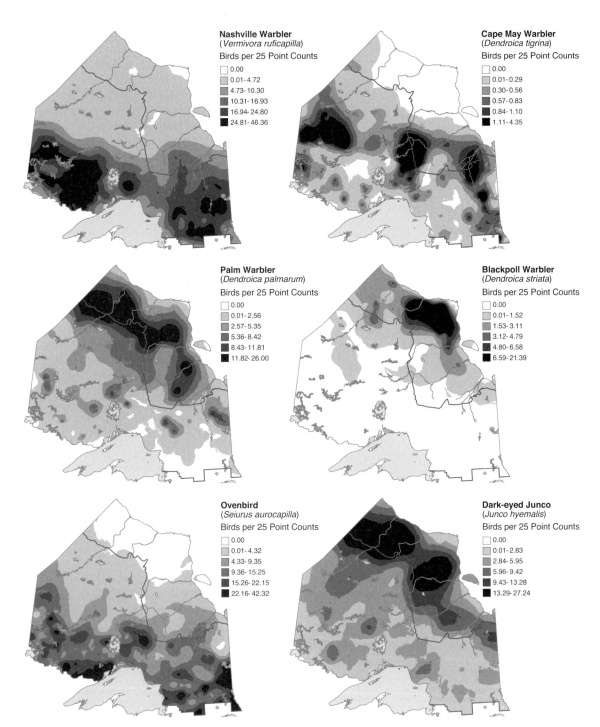

Figure 5.2. Relative abundance maps for six bird species that breed in the boreal region. While all maps use the same relative scale in the legend, note that class breaks are different across species; that is, the actual range of abundance values within a class (and color) is different from one species to the next. Source: Cadman et al. (2007).

Management decisions are often based on incomplete information, or data that are too general for the issue at hand. Bird abundance data, gathered as part of broad-scale atlases, help to define more precisely the areas that are most important for a species, and can therefore help to define priorities for conservation and management.

Continental range maps are useful for "big picture" visualizations of bird distribution, and even for broad-scale assessment of proportional responsibility within countries or provinces (e.g., Blancher et al. 2006), but their precision is inadequate for finer-scale analyses. Atlas data (both quantitative and presence data) can help to validate and refine broad-scale range maps and other occurrence data that may be gathered through a variety of bird monitoring programs.

Abundance sampling at discrete point count locations establishes baseline information on bird abundance that can be gathered at the same locations and using the same methodology again in future atlas projects, so that changes in bird populations can be quantified. Similar types of analyses can be done with presence data, but these are complicated by the non-standard nature of area search methods; thus, the effect of variable survey effort must be addressed (see discussion in Cadman et al. 2007). Quantitative analysis of changes in bird populations over time, and unraveling the causes of these changes, has obvious application to environmental management and policy.

Quantitative data from point counts can be used to generate population estimates for birds within biophysical regions, Bird Conservation Regions (BCRs), political jurisdictions, and so on. This information can be leveraged by the relevant bird groups working under the umbrella of the North American Bird Conservation Initiative (NABCI) to refine species assessment scores and to set population objectives within BCRs.

Finally, and perhaps most interestingly, quantitative data from point counts can be used to delimit landscapes that do a better job of conserving bird populations. By developing spatially explicit models that weigh the needs of a variety of priority bird species, landscape ecologists can define areas on the landscape that represent "best bets" for bird conservation and best value for conservation dollar. Maps of relative abundance of birds therefore take BCR plans to the next practical step by identifying specific areas of interest on the ground.

Abundance information can also be modeled to assess the adequacy of protected areas networks, such as national parks, national wildlife areas, and others. Such networks often have as their mandate the ambitious goal of protecting ecological integrity within and surrounding site boundaries. Clearly, protected areas alone can never hope to maintain populations of most bird species. Relative abundance maps can provide a means to evaluate whether protected areas networks facilitate the long-term persistence of bird populations and to determine if additional protected areas or planning tools are needed.

Abundance data also offer the opportunity to identify formal Important Bird Areas as defined by BirdLife International, a partnership of NGOs in more than 100 countries dedicated to the conservation of the world's birds. Using standardized methods and criteria, BirdLife partners have identified thousands of IBAs worldwide. In Canada, approximately 600 IBAs have been identified to date by BirdLife co-partners Bird Studies Canada and Nature Canada (Lockwood et al. 2005).

Globally, few IBAs have been identified in the boreal biome, and virtually none have been designated in the Canadian boreal forest region. This results from several factors. First, and fundamentally, the IBA concept and criteria do not work well for identifying sites in "frontier" situations. The criteria are biased toward remnant habitats located within settled landscapes. Ironically, identifying IBAs in the boreal forest region will be far easier after the region has been developed and fragmented. Criteria also tend to be biased toward situations where birds are concentrated—a concept that simply does not apply in the breeding season to landbirds in general.

Frontier environments, because of their remoteness and the difficulty of monitoring bird populations, lack sufficient information concerning the distribution and abundance of birds. An added challenge in the boreal forest region is the ecologically dynamic nature of the region—fires, floods, pest outbreaks, and other disturbances have a strong (and ephemeral) influence on bird populations. Superimposed on this natural variability are anthropogenic disturbances associated with established oil and gas leases and forestry concessions. Whether Important Bird Areas, the boundaries of which are expected to remain relatively static over time, can be practically identified within such a naturally dynamic system is an open question. The development of transferable methodological approaches to address these types of situations is therefore highly valuable to the wider BirdLife partnership. New relative abundance data fill an information gap in the boreal forest region and present the possibility of addressing this issue.

LIMITATIONS AND CHALLENGES

Point count censuses, by their very nature, do not sample the full suite of birds occurring in an area. Point counts in the present study were conducted primarily along roadsides, within five hours after sunrise. This meant that birds not active at this time of day were typically not sampled well, for example, raptors, owls, crepuscular species, and so on. Also undersampled were inconspicuous or rare species, colonial species, and/or species with localized and specialized habitat requirements. As a general rule, waterbirds were also not picked up well by the predominantly on-road sampling scheme. In all of these cases, the distribution maps derived from atlas presence data (and from specialized atlas surveys and external databases) proved invaluable, underscoring the importance of a multipronged approach to atlas project design.

A second challenge relates to the statistical properties of the data: even with reasonable sample sizes, kriging is not advisable or possible because of poor model fit. Geostatistical models such as kriging operate on the assumption that a non-linear, distance-decay relationship exists among observations. In other words, nearby observations are assumed to be closer in their values than observations that are farther away. If such a relationship does not exist within the data for a given species (i.e., values appear to be totally random), the model will either (a) not execute or (b) execute, but with a high degree of error and a map with little credibility. In almost all cases, it is immediately obvious to the experienced analyst whether model results and maps are valid or not. For the Ontario Atlas, model results and maps were carefully reviewed by a team of geostatistical analysts and ornithologists. Many maps were deemed unfit for publication due to poor model fit. In addition, the team used expert opinion to determine whether each map made reasonable sense for what we know of the biology of the species in question, and whether the map would add value to the atlas publication. The entire process can, therefore, be a mix of art and science.

However, the conservative approach taken in Ontario ensured that only those maps with a high degree of confidence were published.

In aggregate, these factors limit the number of species for which relative abundance maps can be constructed and accepted. In the Ontario atlas, 279 species were detected on at least one point count; we were able to produce reasonable abundance maps for 130 of these. This result meets pre-project expectations, but the absence of a large number of species nevertheless does constrain post-project analysis and modeling efforts.

A final concern is that point count coverage in the far north beyond the road system remains sparse. Confidence in interpolated predictions is therefore lower in this area relative to the southern boreal forest region. Conspicuous blobs of high (or low) relative abundance, based on a small sample size, may be overstated and are not to be taken as exact. Rather, the maps in this region are intended to act as tools for regional-level planning and to highlight broad patterns of abundance and areas of interest, rather than to provide parcel-level detail. Interpretations and management recommendations focused on specific sites within this area should, therefore, proceed with some degree of caution. The precision of the maps could be improved with more complete sampling of this remote region. The architects of the next Ontario atlas (2021–2025) will no doubt seek to improve coverage in this challenging region. In the meantime, the boreal forest region will continue to change and diverge from the snapshot provided by the second Ontario atlas. In the intervening years, areas of high conservation interest, as well as areas with gaps in coverage, could be selected for targeted research and monitoring.

Atlas projects can be monumental endeavors. They are expensive. They may take seven or eight years to complete from the initial planning stages to the publication of a book or website. Importantly, they motivate, educate, and inspire thousands of volunteer participants. They also focus the attention of governments, nongovernmental organizations, and industry partners on an issue of mutual interest, with an aim to produce credible, reliable information that can improve land management decisions. Relative abundance data add considerable value to atlas projects. In Ontario's case, abundance data provide a start toward improved understanding of the avifauna of the boreal forest region. These data have myriad conservation applications, many of which are already being implemented.

For example, improved population estimates for Bird Conservation Regions and other planning units can help to set broad-level conservation targets. Abundance maps define more precisely where a species occurs in the greatest numbers, and thus help to refine priorities for conservation at a scale that is meaningful to land managers. Complex modeling procedures can be used with abundance data to assess land management scenarios, to evaluate the effectiveness of protected areas networks and, potentially, to identify formal Important Bird Areas. These data will continue to be examined for many years to come and will undoubtedly spawn a great deal of research and improved management, which will enhance bird conservation in the province.

ACKNOWLEDGMENTS

This paper reports on the end result of a long, complex project that involved intensive planning and management on the part of a great many individuals. The author sat on several technical and scientific committees that guided the project and thanks the members of those committees for their insights and expertise, the fruits of which are reflected in this paper. The author also thanks the dedicated project participants who visited more than 60,000 point count locations during the course of the five-year study; without their efforts our goals would not have been realized. The author gratefully acknowledges ESRI Canada for a donation of GIS software and continuing support, and thanks the various financial supporters of the project.

LITERATURE CITED

Blancher, P. J., B. Jacobs, A. R. Couturier, C. J. Beardmore, R. Dettmers, E. H. Dunn, W. Easton, E. E. Iñigo-Elias, T. D. Rich, K. V. Rosenberg and J. M. Ruth. 2006. Making connections for bird conservation: linking states, provinces and territories to important wintering and breeding grounds. Partners in Flight Technical Series No. 4. <http://www.partnersinflight.org/pubs/ts/04-Connections>.

Blancher, P., and J. Wells. 2005. The boreal forest region: North America's bird nursery. Report commissioned by the Boreal Songbird Initiative and the Canadian Boreal Initiative.

Cadman, M. D., P. F. J. Eagles, and F. M. Helleiner (editors). 1987. Atlas of the breeding birds of Ontario. University of Waterloo Press, Waterloo, ON.

Cadman, M. D., D. A. Sutherland, G. G. Beck, D. Lepage, and A. R. Couturier (editors). 2007. Atlas of the breeding birds of Ontario, 2001–2005. Bird Studies Canada, Environment Canada, Ontario Field Ornithologists, Ontario Ministry of Natural Resources, and Ontario Nature, Toronto, ON.

Canadian Boreal Initiative. 2005. The Boreal in the balance: securing the future of Canada's boreal region. Canadian Boreal Initiative, Ottawa, ON.

Dunn, A. M., and M. A. Weston. 2008. A review of terrestrial bird atlases of the world and their application. Emu 108:42–67.

Lockwood, A., A. R. Couturier, and L. S. Wren. 2005. From the tundra to Tierra del Fuego: protecting key sites for birds in Canada and throughout the western hemisphere. Biodiversity 6:40–48.

Pomerey, D., H. Tushabe, and R. Cowser. 2008. Bird atlases—how useful are they for conservation? Bird Conservation International 18:S211–S222.

CHAPTER SIX

The Boreal Landbird Component of Migrant Bird Communities in Eastern North America

Adrienne J. Leppold and Robert S. Mulvihill

Abstract. The vast and remote boreal forest supports nearly 50% of North America's bird species, some of which appear to be in decline and the majority of which are not well monitored. In this study, we provide evidence based on a geographic information system (GIS) analysis of >300,000 banding records that (particularly in fall) the migrations of birds that breed in the boreal forest region from Alaska to the Canadian Maritimes are not simple north–south movements. Rather, they take the form of a large-scale funneling southeast across the Great Lakes and/or southwest along the Appalachians. Boreal birds of some 50 species make up 50% or more of the migrants caught at banding stations located within and near an area of the mid-Appalachian mountains that is about 350 km across. For example, at the Powdermill Avian Research Center in the mountains of southwestern Pennsylvania, 32 species of boreal songbirds comprised 50% of fall captures over a 15-year period. At Allegheny Front Migration Observatory, in the mountains of northeastern West Virginia, 35 species of boreal songbirds comprised 62% of captures over the same period. We propose that the apparent funneling of migrants from across an expansive boreal breeding area through a comparatively narrow "neck" creates an exceptional opportunity for data from coordinated fall migration banding to be used in the monitoring of many species for which other methods are inadequate. Furthermore, it suggests that states within the mid-Appalachian region have a high responsibility for the conservation of boreal landbird migrants.

Key Words: boreal birds, eastern North America, migration monitoring network, migratory funneling, population monitoring, Powdermill Avian Research Center.

The boreal forest is one of the world's largest intact forest ecosystems, spanning 6,000 kilometers (3,500 miles) across Alaska and Canada and 20 degrees of latitude (50°–70°N). Nearly half of all North American birds rely on the boreal forest, especially during the breeding season. More than 1.5 billion landbirds are estimated to breed in the boreal forest region, some of which may be in serious decline (Blancher and Wells 1995, National Audubon Society 2002, Sauer et al. 2007).

In understanding more about populations of these birds, researchers have historically looked to the Breeding Bird Survey (BBS), various breeding bird atlases, and Christmas Bird Count (CBC)

Leppold, A. J., and R. S. Mulvihill. 2011. The boreal landbird component of migrant bird communities in eastern North America. Pp. 73–84 *in* J. V. Wells (editor). Boreal birds of North America: a hemispheric view of their conservation links and significance. Studies in Avian Biology (no. 41), University of California Press, Berkeley, CA.

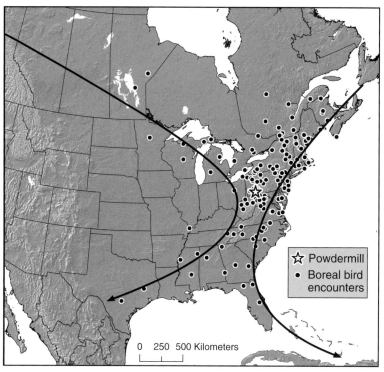

Figure 6.1. Map showing foreign recoveries and foreign encounters of boreal birds for the Powdermill banding program, 1961–2004. The curved lines, drawn by eye, describe the pattern of southward movement of boreal landbird migrants hypothesized in this paper.

data sets. While these have provided a means to estimate some populations, routes and counts have been biased to the southern part of the boreal forest, where the land is more accessible and where observers are much more numerous. Consequently, the remoteness and inaccessibility of much of the northern part of the region have made it a true *terra incognita* with respect to bird population trends. Sample sizes are low, and estimates of bird numbers for many species are imprecise (Bart et al. 2004). Thus, the bird conservation community needs additional methods of monitoring populations of boreal forest–nesting birds in order to detect and investigate possible causes of declining trends (e.g., Dunn et al. 2005).

The Powdermill Avian Research Center (PARC), located in the mountains of southwestern Pennsylvania (40.05°N, 79.16°W), has been the site of a large-volume bird banding program every year since 1961. Species and subspecies composition, as well as recoveries stemming from the Powdermill banding program, provide some evidence that birds originating from boreal forest as far northwest as Alaska and across the Canadian Maritimes, concentrate in

the mid-Appalachian region of the eastern U.S. during their southward migration. Just 36 species of boreal birds (<20% of 190 species of birds banded) comprise 44% of all spring captures and 49% of fall captures at Powdermill. Overall, boreal birds make up half of the top ten species banded at Powdermill from 1961 to 2008: Dark-eyed Junco (*Junco hyemalis*; $n > 36,000$), White-throated Sparrow (*Zonotrichia albicolis*; $n > 17,600$), Ruby-crowned Kinglet (*Regulus calendula*; $n > 13,500$), Yellow-rumped "Myrtle" Warbler (*Dendroica coronata*; $n > 13,200$), and Magnolia Warbler (*D. magnolia*; $n > 12,500$).

A map of band recoveries of boreal species encountered at Powdermill (Fig. 6.1) hints at movements of birds to and from the northwest, across the Great Lakes region. The two encounters farthest to the north and west of Powdermill—a Dark-eyed Junco and a Yellow-rumped Warbler, respectively—suggest a migration trajectory for Powdermill banded birds that can be extended to Alaska. In fact, we regularly catch individuals ascribable to the large Alaskan subspecies of Yellow-rumped Warbler, *D. c. hooveri* (Dunn and Garrett 1997) (144 out of 3,809, or 3.8% of

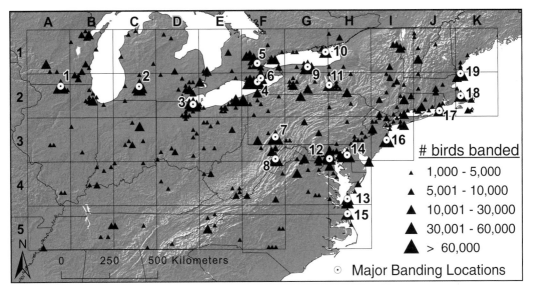

Figure 6.2. Map showing source locations and approximate sample sizes from data used in this study. Major banding sites (i.e., locations contributing >30,000 records) are labeled as follows: 1(Sand Bluff Banding Station), 2 (Kalamazoo Nature Center), 3 (Black Swamp Bird Observatory), 4 (Long Point Bird Observatory), 5 (Ausable Bird Observatory), 6 (Haldimand Bird Observatory/Selkirk), 7 (Powdermill Avian Research Center), 8 (Allegheny Front Migration Observatory), 9 (Braddock Bay Bird Observatory), 10 (Prince Edward Point Bird Observatory), 11 (Kestrel Haven Avian Migration Observatory), 12 (Patuxent Bird Banding Station), 13 (Kiptopeke Bird Banding Station), 14 (Chino Farms Banding Station), 15 (Back Bay Banding Station), 16 (Island Beach), 17 (Block Island Banding Station), 18 (Manomet Bird Observatory), 19 (Shoals Marine Lab/Appledore Island).

birds from 1995 to 2004; birds of this race were not consistently noted prior to 1995), and also the "Cassiar's" Dark-eyed Junco, *J. h. cismontanus* (Leberman 1976) (353 out of 11,619, or 3.0% of birds from 1989 to 2004), whose range extends from the Yukon Territory to central Alberta.

The purpose of this study was to determine if the large, funnel-shaped pattern of migration suggested by Powdermill's data is, in fact, supported by data from other banding stations. That is, do banding data collected throughout the eastern U.S. support the hypothesis that migration of many boreal birds is not a simple north–south movement along a broad longitudinal front, but rather a strongly funneled (especially from the northwest) movement that results in a geographical concentration of boreal birds along the Appalachian Mountains in the mid-latitudes of eastern North America. If so, this would open up the possibility that data from a network of banding stations strategically located near the neck of this hypothesized migratory funnel could be used to assess continental population trends for many boreal species that are otherwise difficult to monitor via standard migration monitoring approaches (e.g., Dunn et al. 1997, Hussell and Ralph 2005).

METHODS

In order to assess the overall distribution of boreal birds at migration banding stations in the eastern U.S., we obtained all original landbird banding records from the Bird Banding Laboratory spanning a 15-year period from 1989 to 2004. Our study area encompassed the mid-latitudes of eastern North America (i.e., 35–45°N and 70–90°W). Because we were specifically interested in assessing the boreal bird composition of migrant communities, only data from the spring (March through May) and fall (August through November) migration seasons were analyzed. These cutoff dates were chosen to eliminate the majority of resident bird captures from the sample and the majority of birds caught as a result of target netting (e.g., baited at feeders or attracted using audio lures). We analyzed a total of 3,544,376 banding records for this study. The banding records themselves were represented by >2,000 unique latitude-longitude coordinates within our study area. However, data from 275 discrete banding locations where >1,000 birds were banded during the 15-year period contributed the vast majority (90%) of the banding data analyzed for this study, and 25 major banding locations (>30,000 birds during the study period) accounted for 50% of the banding records analyzed for this study (Fig. 6.2).

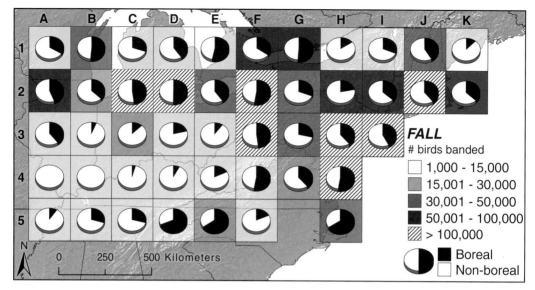

Figure 6.3. Map showing the total number of birds banded in fall (1989–2004) and the percent boreal/not boreal in each of 46 two-degree latitude–longitude grid-blocks covering the mid-latitudes of eastern North America (i.e., the study area for this paper).

We defined as "boreal" 44 species and four distinctive subspecies having an estimated 50% or more of their global breeding population occurring in the boreal forest region of North America, based on Blancher and Wells (2005) (Appendix 6.1). Banding records were summarized for a total of 48 two-degree lat-long blocks and overlaid graphically on a map of our study area. We deleted two of the 48 grid-blocks (J3 and G5) from the study because of small sample sizes (i.e., fewer than 5,000 birds banded in the 15-year period). For all of our spatial analyses, we used ArcGIS 9.2.

To assess geographic concentrations of boreal birds at migration banding stations, we examined the total numbers and percentages of boreal migrants per grid-block for each season. Because the majority of banding records (60%) were from the fall season, and overall patterns proved similar between seasons (although less pronounced in spring), we present detailed data for fall only. Prior to this study, we had computed linear regressions of capture rates (birds per 100 net-hours) on year for selected fall migrants banded at Powdermill from 1962 to 2001. We compared these trends with trends calculated from data submitted by cooperating Canadian Migration Monitoring Network (CMMN) stations, specifically Long Point Bird Observatory (LPBO), the only network station whose banding program spans a time period equivalent to that of Powdermill. The CMMN uses a combination of banding totals and one or more

other migration counts to compute daily estimated totals and analyzes the mean seasonal estimated total using non-linear trend analyses (Hussell and Ralph 2005). Due to differences in methodology, comparisons between CMMN and Powdermill trends in this study are meant to be provisional, that is, simply suggestive of similarities or differences.

RESULTS

Overall, 917,450 (43%) of 2,144,057 fall banding records were boreal birds. The average percentage of boreal birds across all 2° lat-long grids was 33% (range, 1–69%; Appendix 6.2). Highest percentages (≥ca. 50%) of boreal birds occurred in grids adjacent to the shores of the Great Lakes (B1, C2, D2, E1, F2, G1), along the Appalachian Mountains from southwestern Pennsylvania south (F3, F4, D5, E5), and along the Atlantic coast from Delaware south (H4, H5) (Fig. 6.3; Appendix 6.2). Conversely, low average percentages (≤ca. 20%) occurred in grid-blocks due south of the Great Lakes and west of the mountains (A4, A5, B3, B4, C3, C4, D4, E3, E4), southeast of the mountains away from the coast (F5), along the northern New England coast (K5), and east of Lake Ontario (H1) (Fig. 6.3; Appendix 6.2). The overall directions of migration and areas of concentration, that is, funneling, suggested by our geographic analysis of boreal bird percentages at banding sites throughout the mid-latitudes of eastern North America mirror the shape of the proposed distribution based on band encounters of boreal birds from Powdermill (Fig. 6.1).

In our analysis of long-term fall capture-rate trends for 81 selected species at Powdermill (including 30 boreal species), we found nonsignificant trends for the majority (42 species), increasing trends for 25, and decreasing trends for 14 (a complete list of Powdermill trend analyses can be accessed at http://www.westol.com/~banding/Fall_2001_Trend_Table.htm). Within these trend groups, boreal species contributed most to the declining group (8 species; 57%), followed by the nonsignificant trend group (17 species; 40%), and, last, the increasing trend group (5 species; 20%). Four out of five boreal species with significantly increasing trends at Powdermill also had increasing trends based on CMMN analyses of Long Point Bird Observatory data. Of the eight species showing declines based on Powdermill banding data, two (Blackpoll Warbler, *D. striata*, and Nashville Warbler, *Vermivora ruficapilla*) show the opposite trend at LPBO; four (Least Flycatcher, *Empidonax minimus*; Palm Warbler, *D. palmarum*; Wilson's Warbler, *Wilsonia pusilla*; and Lincoln's Sparrow, *Melospiza lincolni*) show no significant trend at LPBO, and two (Olive-sided Flycatcher, *Contopus cooperi*, and Connecticut Warbler, *Oporornis agilis*) were not represented in any of the CMMN station's statistical analyses.

DISCUSSION

Boreal birds contribute greatly to the abundance and diversity of landbirds migrating through the mid-latitudes of eastern North America, just as they do to wintering bird communities in the southern U.S. and throughout the Neotropics (Robertson et al., this volume, chapter 7). The observed pattern of boreal bird concentrations at banding locations within our study area provide support for the hypothesis, based on long-term observations at a single banding location (PARC), that their fall migrations do not take the form of simple north–south movements across a broad longitudinal front. The large percentage contributions of boreal birds to fall banding samples along the shores of the western Great Lakes are not evident at latitudes immediately to the south. Instead, the pattern indicates that the overall movement of boreal birds is southeast across the Great Lakes and then southwest along the main axis of the Appalachian Mountains. This convergence of

boreal species from both the northwest and northeast toward the Great Lakes is suggested by the spread of banding encounters at Powdermill (Fig. 6.1), as well as by band recoveries for stations in the CMMN (Dunn et al. 2006). Birds of northwestern origin that do not "reorient" from their original southeasterly course upon reaching the Appalachian Mountains eventually will find themselves at or beyond the Atlantic coastline. Evidence suggests that, disproportionately, these are inexperienced hatching year birds with survival probabilities much lower than adults (Ralph 1978, 1981). This, in turn, suggests that data from Atlantic coastal banding stations may not be entirely suitable for population monitoring purposes. Of course, not all boreal birds caught at Atlantic coastal banding stations are inexperienced overshoot migrants—many undoubtedly originate in the northeastern boreal habitats and simply move southward along the Atlantic Coast well east of the Appalachian Mountains.

However, because so many boreal birds do appear to funnel toward the Great Lakes and the mid-Appalachians, banding stations along this route, and especially those closest to the "neck" of the migration funnel (Fig. 6.2; e.g., sites 7 and 8—Powdermill and Allegheny Front), may be particularly well positioned to collect migration data that are useful for monitoring their populations (Dunn 2005). Importantly, results of our study also can be used to identify areas where the establishment of additional major banding stations would be strategic in the context of a mid-latitude migration monitoring network for boreal birds—for example, ridges in southern West Virginia, western Virginia, western North Carolina, and eastern Tennessee. Boreal bird species were well represented, but the overall banding totals were comparatively small in these areas (Fig. 6.3).

In the future, obtaining information about the geographical source areas from which migrant banding samples are drawn—for example, using feather isotopes, genetic data, and/or morphometrics—will greatly increase the population monitoring value of data collected at these sites (e.g., Dunn et al. 2006). In addition, if it can be shown that annual variation in the geographic origin(s) of migrants at individual banding sites is limited, then analysis of data from long-term banding sites like Powdermill Avian Research Center, Long Point Bird Observatory, and Allegheny Front Migration Observatory could potentially be used not only to monitor future population change, but also to provide valuable information on historical trends.

Based on results of our study—and following the recommendations of Carlisle and Ralph (2005)—we conclude that the focused collection and analysis of data from a clustered network of existing and newly established migration banding sites farther south than those within the CMMN, located within and near the mid-Appalachian region where boreal birds appear to become much more concentrated, especially in fall, will prove to be an efficient and reliable means for assessing population trends for boreal birds—in some cases, perhaps the only means for scarce or secretive species, such as Connecticut Warbler and Olive-sided Flycatcher. Importantly, because of well-known limitations and biases associated with the use of mist nets for conducting quantitative bird surveys (Dunn and Ralph 2004), we strongly recommend that banding sites within any such future migration monitoring network all follow published recommendations for the proper use of mist nets in monitoring efforts (Ralph et al. 2004), as well as advice regarding the incorporation at every banding site of at least one additional daily count method for estimating numbers of boreal birds (Dunn et al. 2004).

Finally, we think the findings of this study have implications beyond the possible strategic importance of the mid-Appalachians for monitoring populations of landbirds whose origins are in the expansive and largely inaccessible boreal forest region of North America. The same small neck of a broad migration funnel that provides opportunities for population monitoring may, ironically, constitute a population "bottleneck" for boreal landbirds. This is because the mid-Appalachians currently are experiencing rapidly increasing land use pressure related to extraction of fossil fuels (e.g., coal and natural gas) and the development of renewable wind energy. In short, governmental and nongovernmental agencies working in the region have both high responsibility and significant opportunities for promoting the conservation of boreal landbird migrants.

ACKNOWLEDGMENTS

First and foremost, we thank Jeff Wells of the Boreal Songbird Initiative for inviting us to participate in the Boreal Bird Symposium at the IV North American Ornithological Conference in Veracruz, Mexico, in October 2006. We are grateful to Kathy Klimkiewicz and Danny Bystrak at the USGS Patuxent Bird Banding Laboratory for filling our large data request in a timely manner. We also thank our own database manager, Marilyn Niedermeier, for filling our in-house data requests related to this study. Powdermill GIS Technician Kristin Sesser conducted the spatial analyses for our study and prepared all of the map figures—without her hard work and patience in accomplishing these tasks, we simply could not have completed this study in time for its presentation at the IV NAOC. We acknowledge and greatly appreciate the hard work and data contributed by the hundreds of banders who submitted records to the BBL within the area and for the years analyzed in our study. The Powdermill banding program owes its existence to its founder in 1961, Robert C. Leberman, and the program's success and productivity is thanks largely to him, to the financial support of many private individuals and foundations, and to the helpful efforts of dozens of dedicated volunteers over nearly a half century. Finally, we thank two anonymous reviewers for comments and insights that helped improve the paper.

LITERATURE CITED

Bart, J., K. P. Burnham, E. H. Dunn, C. M. Francis, and C. J. Ralph. 2004. Goals and strategies for estimating trends in landbird abundance. Journal of Wildlife Management 68:611–626.

Blancher, P., and J. V. Wells. 2005. The boreal forest region: North America's bird nursery. Boreal Songbird Initiative and Canadian Boreal Initiative, Ottawa, ON.

Carlisle, J. D., and C. J. Ralph. 2005. Towards the establishment of landbird migration monitoring networks in the United States. USDA Forest Service General Technical Report PSW-GTR-191. USDA Forest Service, Redwood Science Laboratory, Arcata, CA.<http://www.fs.fed.us/psw/publications/documents/psw_gtr191/Asilomar/pdfs/698-700.pdf [last accessed 7 September 2008]>.

Dunn, E. H. 2005. Counting migrants to monitor bird populations: state of the art. USDA Forest Service General Technical Report PSW-GTR-191.

Dunn, E. H., B. L. Altman, J. Bart, C. J. Beardmore, H. Berlanga, P. J. Blancher, G. S. Butcher, D. W. Demarest, R. Dettmers, W. C. Hunter, E. E. Inigo-Elias, A. O. Panjabi, D. N. Pashley, C. J. Ralph, T. D. Rich, K. V. Rosenberg, C. M. Rustay, J. M. Ruth, and T. C. Will. 2005. High priority needs for range-wide monitoring of North American landbirds. Partners in Flight Technical Series No. 2. <http://www.partnersinflight.org/pubs/ts/02-MonitoringNeeds.pdf> (7 September 2008).

Dunn, J. L., and K. L. Garrett. 1997. A field guide to the warblers of North America. The Peterson field guide series. Houghton Mifflin Co., New York, NY.

Dunn, E. H., K. A. Hobson, L. I. Wassenaar, D. J. T. Hussell, and M. L. Allen. 2006. Identification of summer origins of songbirds migrating through southern Canada in autumn. Avian Conservation and Ecology 1(2):4. <http://www.ace-eco.org/vol1/iss2/art4/> (7 September 2008).

Dunn, E. H., D. J. T. Hussell, and R. J. Adams. 1997. Monitoring songbird population change with autumn mist netting. Journal of Wildlife Management 61:389–396.

Dunn, E. H., D. J. T. Hussell, C. M. Francis, and J. D. McCraken. 2004. A comparison of three count methods for monitoring songbird abundance during spring migration: capture, census, and estimated totals. Pp. 116–122 in C. J. Ralph and E. H. Dunn (editors), Monitoring bird populations using mist nets. Studies in Avian Biology No. 29, Cooper Ornithological Society.

Dunn, E. H., and C. J. Ralph. 2004. Use of mist nets as a tool for bird population monitoring. Pp. 1–6 in C. J. Ralph and E. H. Dunn (editors), Monitoring bird populations using mist nets Studies in Avian Biology No. 29, Cooper Ornithological Society.

Hussell, J. T., and C. J. Ralph. 2005. Recommended methods for monitoring change in landbird populations by counting and capturing migrants. North American Bird Bander 30:6–20.

Leberman, R. C. 1976. The birds of the Ligonier Valley. Carnegie Museum Special Publication No. 7. Pittsburgh, PA.

National Audubon Society. 2002. The Christmas bird count historical results <http://www.audubon.org/bird/cbc> (4 October 2007).

Ralph, C. J. 1978. The disorientation and possible fate of young coastal passerine migrants. Bird Banding 39:237–247.

Ralph, C. J. 1981. Age ratios and their possible use in determining routes of passage migrants. Wilson Bulletin 93:164–188.

Ralph, C. J., E. H. Dunn, W. J. Peach, and C. M. Handel. 2004. Recommendations for the use of mist nets for inventory and monitoring of bird populations. Pp. 187–196 in C. J. Ralph and E. H. Dunn (editors), Monitoring bird populations using mist nets. Studies in Avian Biology No. 29, Cooper Ornithological Society.

Sauer, J. R., J. E. Hines, and J. Fallon. 2007. The North American Breeding Bird Survey, results and analysis 1966–2006, version 7.23.2007. USGS Patuxent Wildlife Research Center, Laurel, MD.

Boreal landbird species and subspecies and the total number of each banded from 1989 to 2004 in the mid-latitudes of eastern North America

Common Name	Scientific Name	No. Banding Records
Yellow-bellied Sapsucker	*Sphyrapicus varius*	2,842
Olive-sided Flycatcher	*Contopus cooperi*	195
Yellow-bellied Flycatcher	*Empidonax flaviventris*	10,420
Alder Flycatcher	*Empidonax alnorum*	188
Least Flycatcher	*Empidonax minimus*	20,867
Northern Shrike	*Lanius excubitor*	187
Cassin's Vireo	*Vireo cassinii*	1
Blue-headed Vireo	*Vireo solitarius*	8,507
Philadelphia Vireo	*Vireo philadelphicus*	4,775
Gray Jay	*Perisoreus canadensis*	72
Boreal Chickadee	*Poecile hudsonica*	7
Ruby-crowned Kinglet	*Regulus calendula*	113,993
Gray-cheeked Thrush	*Catharus minimus*	14,677
Swainson's Thrush	*Catharus ustulatus*	66,902
Hermit Thrush	*Catharus guttatus*	57,446
Bohemian Waxwing	*Bombycilla garrulous*	83
Tennessee Warbler	*Vermivora peregrina*	40,378
Orange-crowned Warbler	*Vermivora celata*	2,554
Nashville Warbler	*Vermivora ruficapilla*	26,426
Magnolia Warbler	*Dendroica magnolia*	96,834
Cape May Warbler	*Dendroica tigrina*	12,521
Yellow-rumped "Myrtle" Warbler	*Dendroica coronata coronata*	228,791
Yellow-rumped "Audubon's" Warbler	*Dendroica c. auduboni*	10
Black-throated Green Warbler	*Dendroica virens*	17,327
Blackburnian Warbler	*Dendroica fusca*	5,834
"Western" Palm Warbler	*Dendroica palmarum palmarum*	17,010
"Yellow" Palm Warbler	*Dendroica p. hypochrysea*	4,417
Bay-breasted Warbler	*Dendroica castanea*	9,056
Blackpoll Warbler	*Dendroica striata*	37,739
Black-and-white Warbler	*Mniotilta varia*	22,797
Northern Waterthrush	*Seiurus noveboracensis*	24,367
Connecticut Warbler	*Oporornis agilis*	2,374
Mourning Warbler	*Oporornis philadelphia*	7,800

APPENDIX 6.1 (*continued*)

Common Name	Scientific Name	No. Banding Records
Wilson's Warbler	*Wilsonia pusilla*	18,269
Canada Warbler	*Wilsonia canadensis*	16,238
Clay-colored Sparrow	*Spizella pallida*	134
Le Conte's Sparrow	*Ammodramus leconteii*	23
Fox Sparrow	*Passerella iliaca*	9,880
Lincoln's Sparrow	*Melospiza lincolnii*	14,997
Swamp Sparrow	*Melospiza georgiana*	45,852
White-throated Sparrow	*Zonotrichia albicollis*	210,274
White-crowned Sparrow	*Zonotrichia leucophrys*	17,306
"Eastern" White-crowned Sparrow	*Z. l. leucophrys*	5,954
"Gambel's" White-crowned Sparrow	*Z. l. gambelii*	163
Dark-eyed Junco	*Junco hyemalis*	181,097
Rusty Blackbird	*Euphagus carolinus*	1,350
Pine Grosbeak	*Pinicola enucleator*	45
White-winged Crossbill	*Loxia leucoptera*	35

Sample sizes by map grid from Figure 6.3

Grid ID	Boreal	Not Boreal	Total
A1	3,746	7,680	11,426
A2	22,193	28,731	50,924
A3	1,605	2,562	4,167
A4	33	5,664	5,697
A5	403	3,500	3,903
B1	20,730	19,697	40,427
B2	10,839	19,125	29,964
B3	359	4,697	5,056
B4	92	9,394	9,486
B5	2,104	5,169	7,273
C1	858	2,051	2,909
C2	71,716	78,528	150,244
C3	2,623	16,820	19,443
C4	85	1,697	1,782
C5	1,428	3,581	5,009
D1	3,018	4,662	7,680
D2	69,565	68,891	138,456
D3	3,093	11,215	14,308
D4	160	1,712	1,872
D5	4,193	1,878	6,071
E1	7,577	6,513	14,090
E2	17,201	26,599	43,800
E3	1,089	8,846	9,935
E4	2,238	10,383	12,721
E5	16,642	8,581	25,223
F1	29,738	58,571	88,309
F2	129,631	115,161	244,792
F3	92,675	101,058	193,733
F4	3,999	3,503	7,502
F5	932	3,971	4,903
G1	41,390	39,353	80,473
G2	9,122	21,083	30,205
G3	12,622	32,192	44,814
G4	10,407	17,036	27,443

APPENDIX 6.2 (*continued*)

Grid ID	Boreal	Not Boreal	Total
H1	952	4,902	5,854
H2	13,951	48,510	62,461
H3	57,147	92,747	149,894
H4	67,378	60,194	127,572
H5	22,192	11,249	33,441
I1	4,000	9,156	13,156
I2	19,278	34,689	53,967
I3	57,863	81,182	139,045
J1	12,320	17,993	30,313
J2	38,767	61,514	100,281
K1	421	2,863	3,284
K2	28,630	50,510	79,140

CHAPTER SEVEN

Boreal Migrants in Winter Bird Communities

Bruce A. Robertson, Rich MacDonald, Jeffrey V. Wells,
Peter J. Blancher, and Louis Bevier

Abstract. An estimated 3–5 billion birds migrate south from the boreal forest ecoregions of Alaska and Canada every fall to winter across the Americas. Many of these boreal forest breeding birds become integrated into winter bird communities in Central and South America and the Caribbean, but their significance and ecological roles are only beginning to be understood. We used existing data sets on distribution and relative abundance to explore both the biogeography of wintering boreal forest migrants and their occurrence and relative abundance within wintering bird communities. We report on (1) the proportion of the wintering populations of boreal migrants estimated to occur within countries and ecological regions; (2) the comparative diversity and abundance of boreal migrants across countries and ecoregions; (3) their occurrence, relative abundance, and diversity within local bird communities; and (4) a survey of the ecological roles of wintering boreal migrants.

Species richness, abundance, and density of boreal forest migrants were especially high in Mexico, Central America, and Caribbean countries. Results show that boreal forest migrants of different taxa and from different boreal forest ecoregions and taxonomic groups incorporate themselves into winter bird communities in different regions of Latin America. Our review of the literature demonstrates that boreal forest migrants are commonly the most speciose portion of the migrant community in winter bird communities, that they can numerically dominate the resident bird community, and that they play important ecological roles on their wintering grounds. Consequently, our results support the contention that boreal migrants are significant ecological components of the various communities they inhabit.

Key Words: boreal birds, migratory birds, wintering distribution, wintering ecology.

North America's boreal forest encompasses 5.9 million square kilometers, making up one-quarter of the world's intact forest ecosystems (Bryant et al. 1997, Lee et al. 2006). Nearly half of all North American birds (325 species) breed in the region (Wells and Blancher, this volume, chapter 2). It is estimated that 93% of boreal forest landbirds—that is, between 3 and 5 billion adult and immature birds—migrate from the region every fall (Blancher 2003). Many of these boreal forest–breeding birds become integrated into North

Robertson, B. A., R. MacDonald, J. V. Wells, P. J. Blancher, and L. Bevier. 2011. Boreal migrants in winter bird communities. Pp. 85–94 *in* J. V. Wells (editor). Boreal birds of North America: a hemispheric view of their conservation links and significance. Studies in Avian Biology (no. 41), University of California Press, Berkeley, CA.

American and Neotropical bird communities in Central and South America and throughout the Caribbean. There has been an increasing volume of work investigating these nearctic migrants in specific winter locales, but their significance and ecological roles on a much broader scale are only just beginning to be understood. Such information is needed to gain a more holistic understanding of the ecological importance of boreal species in avian communities during the Northern winter and is a prerequisite to assess the importance of wintering boreal forest migrants in shaping ecosystem processes, food webs, and community structure on large scales.

In this study, we used existing data sets on distribution and relative abundance to explore the winter biogeography of migrants that breed in the boreal forest. We refer to these migrants as "boreal migrants" and investigate their occurrence and relative abundance within wintering bird communities with four primary objectives: (1) to explore the proportion of the wintering populations of boreal migrants estimated to occur within various countries; (2) to assess the comparative diversity and abundance of boreal migrants across countries; (3) to begin to describe the occurrence, relative abundance, and diversity within local bird communities; and (4) to present a survey of the ecological roles of wintering boreal migrants.

METHODS

In this study, we focused on 190 species of landbirds that breed in the boreal forest regions of Canada and winter farther south and for which we had data to estimate the size of the boreal forest breeding population (Blancher 2003). In general, we examined patterns of diversity and abundance of boreal migrants by estimating species richness, bird density, and richness weighted as a percent of global population from the boreal in countries throughout the Neotropics. This analysis considered 33 countries throughout North, South, and Central America, including islands throughout the Caribbean Sea. For ease of analysis, we pooled the remaining Antillean republics into the commonly used categories of "windward" and "leeward" island groups. We overlaid digital range maps created by Ridgely et al. (2003) onto political maps to calculate species richness for each country, simply tallying the number of species whose winter ranges overlapped each

country. The proportion of each species' winter range in individual countries was calculated according to the land area of extant range within each country.

We estimated each species' abundance in wintering areas from the proportion of winter range of that species in the country in question, multiplied by the estimated number of birds migrating out of the boreal region after the breeding season (Blancher 2003). Using data from the North American Breeding Bird survey and Partners in Flight methods for converting relative to absolute abundance (Rich et al. 2004, Blancher et al. 2007), Blancher (2003) estimated that approximately 3–5 billion birds migrate out of the boreal forest region each year.

To map the geographic patterns of species richness and density, we overlaid a grid of 1° latitude × 1° longitude blocks over the Ridgely et al. (2003) digital range maps. Species richness and abundance were then estimated per degree block in the same manner as for countries above, but based on area of range and land area within individual degree blocks. That is, for density maps, numbers of birds of each migrant species were allocated to degree blocks based on the proportion of their wintering range in the 1° × 1° degree block.

Maps of species richness and abundance give unequal weight to each species; that is, species with larger winter ranges have more influence on richness maps than species with restricted range, and common birds have more influence on abundance maps than rare species. We created a third type of map ("weighted richness maps") that weights each species equally, by summing across species the proportion of winter range of each species found there.

All of our winter density and abundance calculations and maps assume a uniform distribution of individuals throughout a species' winter range. This assumption may approximate the distribution of individuals for some species, but will be generally unrealistic for species with highly clumped distributions. Consequently, our winter distribution maps and density and abundance estimates should be considered coarse-scale approximations.

Because populations may mix to different degrees during the breeding and non-breeding season, the strength of connectivity between different regions is also important in assessing threats and for designing recovery plans for declining species (Webster et al. 2002), but also for understanding whether regional populations of boreal migrants integrate into the ecological communities in different regions during winter. Bird Conservation Regions (BCRs) are ecologically distinct regions in North America with similar bird communities, habitats, and resource management issues developed by the North American Bird Conservation Initiative (U.S. NABCI Committee 2000). North America's boreal forest region is divided into four bird conservation regions (see Fig. 2.1, p. 8; U.S. NABCI Committee 2000) that allow a coarse resolution for evaluating the relative strength of migratory connectivity between boreal and wintering regions. To examine variation in the importance of wintering regions to boreal migrants from different boreal forest regions, we created maps of wintering species richness for species breeding in each of the four bird conservation regions. To illustrate geographic variation in the use of value of different regions by boreal migrants of different taxonomic groups, we created maps of wintering species richness for four families of boreal birds: sparrows (Emberizidae), thrushes (Turdidae), wood warblers (Parulidae), and flycatchers (Tyrannidae).

RESULTS

Species richness of wintering boreal migrants was greatest in Mexico (115 species), followed by Guatemala (70) and the United States (66) (Table 7.1). Central American countries were particularly diverse in boreal migrants, representing seven of the top nine most speciose countries. At the $1° \times 1°$ scale, regions of highest boreal species richness were also concentrated in the United States and Mexico and generally appeared higher in coastal regions (Fig. 7.1A). In terms of sheer numbers of boreal migrants, the United States and Mexico hosted the greatest numbers of individuals, with an estimated 1.1 billion and 680 million, respectively (Table 7.1). In contrast to patterns of species richness, the next five most important countries in terms of abundance were located in South America (Brazil, Colombia, Venezuela, Peru, and Bolivia). Density patterns at the 1° latitude by 1° longitude scale paralleled those of species richness with the United States, Mexico, and Central American countries exhibiting the

TABLE 7.1

Estimated species richness and abundance of boreal migratory birds by country in winter

The species pool included 190 landbird species detected on Breeding Bird Survey (BBS) routes in the Canadian boreal. Abundance estimates are based on BBS relative abundance data and are generally conservative, relative to estimates derived from Breeding Bird Census densities.

Country	No. Boreal Migrant Species	Millions of Migrants
Mexico	115	680
Guatemala	70	34
U.S.A.	66	1,150
Honduras	63	30
Nicaragua	58	31
El Salvador	55	6
Belize	54	7
Costa Rica	54	12
Panama	53	15
Columbia	53	110
Venezuela	46	62
Brazil	42	200
Cuba	40	22
Ecuador	40	15
Peru	40	52
Bahamas	36	3
Guyana	32	8
Jamaica	31	2
Bolivia	31	39
Cayman Islands	30	<1
Haiti	26	4
Dominican Republic	26	7
Puerto Rico	26	1
Argentina	24	34
Paraguay	24	11
Trinidad & Tobago	24	<1
Windward Islands	23	<1
Leeward Islands	22	<1
Suriname	22	4
French Guiana	19	2
Chile	18	4
Uruguay	17	2
Canada (non-Boreal)	4	<1
Bermuda	1	<1
Falkland Islands	0	0

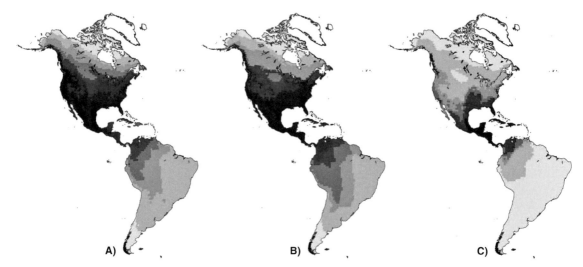

Figure 7.1. Species richness (A), density (B), and density weighted by the percent of the global population that breeds in Canada's boreal forest (C) for boreal birds in winter. Darker regions represent higher species richness (A), density (B) or weighted richness (C) at 1° latitude by 1° longitude resolution.

greatest densities of boreal migrants (Fig. 7.1B). When weighted by percentage of the global population, richness was greatest in southern Mexico and countries of Central America and the Caribbean (Fig. 7.1C).

Species richness associated with the northwestern interior forest (BCR 4) was generally greatest in coastal regions of western Canada, the United States, and Mexico (Fig. 7.2A). In contrast, richness of migrants originating in the boreal taiga plains (BCR 6) was greatest throughout the landmasses between the equator and 30–35° north latitude (Fig. 7.2B). Species from the taiga shield and Hudson plains were highly concentrated in the southern Great Plains, the southeastern United States, and Caribbean islands (Fig. 7.2C). Highest richness of boreal migrants associated with the boreal softwood shield tended to be concentrated further south and were generally higher in northern South America than those of other BCRs (Fig. 7.2D).

Sparrows and thrushes breeding in the boreal forest region were concentrated in the southern U.S. and Mexico, but thrushes had important segments of their populations extending into central and South America (Fig. 7.3A, B). Thrush diversity was particularly concentrated along the west coast of the United States, while sparrow diversity peaked in the southern Great Plains. Boreal flycatchers and wood warblers exhibited primarily neotropical distributions during winter (Fig. 7.3C, D).

DISCUSSION

Geographic Patterns of Boreal Migrant Species Richness, Abundance, and Density

Boreal migrant diversity exhibits important peaks throughout the southern temperate and neotropical regions, but were generally more concentrated north of the equator. As a group, the United States, Central American countries, and Mexico incorporate the most species into their wintering bird communities. Even so, these patterns of species richness are generally paralleled at a finer scale of geographic resolution (Fig. 7.1A), and these same areas appear to incorporate large numbers of boreal birds into their ecological communities (Fig. 7.1B). When simultaneously considering the numbers of species and individuals entering wintering communities, Central America, Mexico, and the Caribbean are especially important to boreal migrant bird communities (Fig. 7.1C).

We found latitudinal and longitudinal links in species richness between discrete breeding and wintering areas. The greatest diversity of boreal migrant species of more westerly origin (northwestern interior forest) occurred during

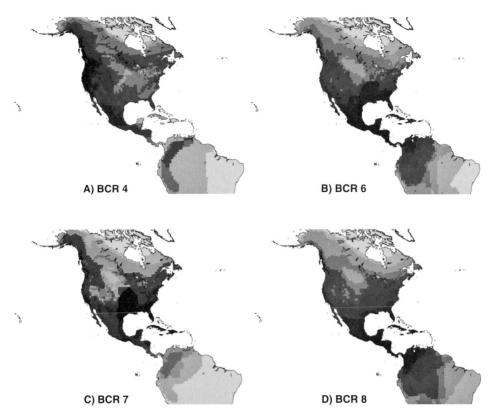

Figure 7.2. Species richness of boreal migrants in winter, weighted by percent of global population that breeds in four NABCI Bird Conservation Regions (BCRs): (A) 4, northwestern interior forest; (B) 6, boreal Taiga Plains; (C) 7, Taiga Shield and Hudson Plains; and (D) 8, boreal softwood shield. Darker regions indicate higher weighted richness, calculated at the 1° latitude by 1° longitude scale.

winter along the west coast of North America. In general, winter species richness of boreal migrants associated with breeding areas in the more northerly taiga shield and Hudson plains (BCR 7) was more northerly distributed than those breeding in the more southerly boreal softwood shield (BCR 8; Fig. 7.2C, D). Boreal migrants associated with the boreal taiga plains were particularly rich in the southern Great Plains (BCR 6), southeastern United States, and Caribbean. The distribution of sparrow species richness explained southern Great Plains richness patterns associated with BCR 6. Thrushes, flycatchers, and warblers had important components of their distributions extending into South America, with flycatchers and warblers exhibiting predominantly neotropical distributions (Fig. 7.3). These results indicate important links between the bird communities of particular boreal forest and neotropical regions.

Occurrence, Relative Abundance, and Diversity within Local Neotropical Bird Communities

Numerous studies have included discussion of the occurrence, relative abundance, and diversity of boreal migrants within bird communities during the temperate winter. Estimates of the percentage of bird communities composed of wintering nearctic migrants can vary according to survey methods, habitat, and region. In neotropical locations, estimates commonly center around 35% (e.g., Robbins et al. 1992, Wunderle and Waide 1993) and range from 29% (Dominican Republic: Terborgh and Faaborg 1980, Latta et al. 2003) to as high as 39% (Cuba: Wallace et al. 1996). In more northerly locations, the percentage is more variable (e.g., Georgia [14%]: White et al. 1996; Nebraska [14%]: Poague et al. 2000; Oregon [30%]: Reinkensmeyer et al. 2008). Boreal migrants typically represent a significant portion of the total. During 18 years of monitoring in Guánica, Puerto Rico, Faaborg and

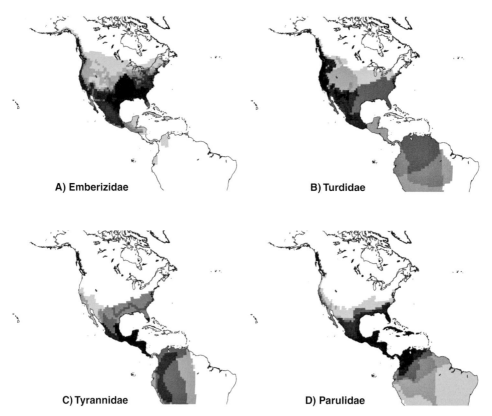

Figure 7.3. Species richness of four families of boreal migratory birds in winter, weighted by the percent of global population that breeds in Canada's boreal forest: (A) sparrows (Emberizidae); (B) thrushes (Turdidae); (C) tyrant flycatchers (Tyrannidae); and (D) wood warblers (Parulidae). Darker regions indicate higher weighted richness at a geographic resolution of 1° latitude by 1° longitude.

Arendt (1992) found that six of the ten Nearctic–neotropical migrants detected were boreal species. In the U.S. Virgin Islands, Askins et al. (1992) found nine out of 16 (56%) neotropical migrants detected during point counts were boreal migrants, accounting for nearly a third of all individuals detected. On the Caribbean island of Hispanola, 54% of wintering migrants were boreal-breeding species (Latta et al. 2003). Robbins et al. (1992) found that six of the 14 Nearctic–neotropical migratory species detected in agricultural and forest sites in Puerto Rico, Jamaica, Belize, and Costa Rica in midwinter were boreal migrants. A single boreal migrant (Black-and-white Warbler, *Mniotilta varia*) represented the second highest number of captures of any migrants. Boreal species were among the most common bird species of any kind exploiting milpa agriculture areas, pasture, acahual, and forest in the Yucatán Peninsula during winter (Lynch 1992). In his studies of wintering landbirds throughout western Mexico, Hutto (1992) found boreal migrants to be a high proportion of the total species pool in eight of the 14 (57%)

vegetation types surveyed. Wilson's Warbler (*Wilsonia pusilla*) was four times more abundant in the two-zone (deciduous and cloud) forest than the next most abundant resident and was one of two boreal migrants represented in this vegetation type. In the two-zone (thornforest and pine–oak–fir forest) vegetation type, of the ten species observed, Yellow-rumped Warbler (*Dendroica coronata*) was the only migrant and was more than twice as abundant as the next most common resident. Together, these results illustrate that boreal migrants commonly represent a sizable fraction of the pool of wintering migratory birds and can be numerically dominant members of wintering bird communities.

Functional Importance of Boreal Migrants to Winter Communities

Migratory birds are components of the ecological communities on their wintering grounds for several months during each year, comprising anywhere from 4% to 50% of the species composition of some feeding guilds (Hutto 1980, Stiles 1980). Migrants, like other species, are part of the food web and the energy and productivity pathways, and so affect other community members through indirect ecological interactions. Many authors (e.g., Slud 1960, Rappole 1995) point out that migrants tend to ecologically complement the resident community, frequently fitting into seemingly unoccupied niches.

Some boreal migrant species have important direct and specialized relationships as pollinators, seed dispersers, and predators with members of the communities they overwinter in. Though most boreal migrant landbirds are insectivorous in North America, many migratory warblers (Parulidae, e.g., Bay-breasted Warbler), thrushes (Turdidae), vireos (Vireonidae), and flycatchers (Tyrannidae) shift to a diet consisting largely or entirely of fruit (Morton 1971, Blake and Loiselle 1992). Migrant abundance is often higher in secondary forest habitats than in primary forest (Karr 1976, Hutto 1980, Martin 1985), a fact commonly attributed to the greater abundance of smaller fruits in these habitats (Martin 1985). Some tropical trees even time fruit maturation to coincide with the migration of neotropical migratory birds (Morton 1971, Howe and De Steven 1979, Levey 1988). A high density of migrants in fruit-rich secondary habitats (e.g., 54% of individuals; Reid et al. 2008) could profoundly affect the trajectory and pattern of forest restoration in degraded landscapes if migrants are effective at dispersing seeds between patches differing in successional stage. In this way, boreal migrants could play a key role not only in maintaining populations of fruiting trees and plants, but indirectly on the population dynamics of other plants and animals that are dependent on tropical forest.

Tennessee Warbler (*Vermivora peregrina*) is such an important pollinator of the tropical forest vine *Comretum fructicosum*, from which the bird obtains an important nectar food source, that Morton (1980) even hypothesized a co-evolved relationship between these two species. Cape May Warbler (*Dendroica tigrina*) plays a similar ecological role as plant pollinator in the Greater Antilles (Baltz and Latta 1998). Our analyses suggest that much of the world's estimated population of 3.2 million Cape May Warblers (Rich et al. 2004) winter in Cuba and the Dominican Republic, leading to questions about how reduced abundance or absence of this species would affect populations of the tropical plant species they pollinate and the fauna that subsist on those plants. A hurricane in the Bahamas virtually eliminated two bird species that pollinated the shrub *Pavonia bahamensis*, resulting in a decline in fruit set of 74% (Rathcke 2000).

Boreal bird species are important in controlling outbreaks of defoliating insects on their breeding grounds (Takegawa and Garton 1984) and may play an important role in regulating insect populations on their non-breeding grounds as well. In the two-zone (cloud forest and pine–oak–fir forest) vegetation type, Ruby-crowned Kinglets (*Regulus calendula*) are the most frequently detected bird species during winter (Hutto 1992). In this particular vegetation type, the highest migrant count was only 22% of the kinglet count; the highest resident species tallied 51% of the kinglet count. Such a high number of kinglets suggests that this species may play an important functional role in this bird community and the ecosystem in general. Several studies have shown that birds reduce insect densities (Holmes et al. 1979, Atlegrim 1989, reviewed by Whelan et al. 2008), especially when insect populations are at either low or more typical levels (Crawford and Jennings 1989, Holmes 1990, Torgerson et al. 1990). In general, the ecological impact of migrants on tropical communities has received relatively little study, but these few examples highlight important ecological roles for boreal migrants. As such, boreal migrants are significant components of the ecological communities they inhabit during the temperate winter, not a superfluous and tangential element (Rappole 1995).

Results of this study and literature survey suggest that the migratory birds of one ecoregion—the vast boreal forest region of Canada—may have important ecological impacts on other, distant regions. The frequency and abundance of overwintering boreal species in bird communities along with their potentially diverse ecological services and interspecific interactions suggest that shifts in their abundance or occurrence may have significant ecological consequences on the residents of their winter communities, perhaps even with the potential to alter community composition or ecosystem function. Research into the population consequences of habitat choice and ecology of overwintering migrants is decreasing (Faaborg et al. 2009) and remains a low funding priority throughout Latin America and the Caribbean (Castro and Locker 2000). More information is needed about the functional and ecological importance of boreal migratory birds to the biological communities into which they assimilate during the temperate winter.

ACKNOWLEDGMENTS

Thanks to the Boreal Songbird Initiative, Bird Studies Canada, the Canadian Boreal Initiative, Partners in Flight, and Michigan State University for funding and logistical support.

LITERATURE CITED

Askins, R. A., D. N. Ewert, and R. L. Norton. 1992. Abundance of wintering migrants in fragmented and continuous forests in the U.S. Virgin Islands. Pp. 197–206 in J. M. Hagan III and D. W. Johnston (editors), Ecology and conservation of neotropical migrant landbirds. Smithsonian Institution Press, Washington, DC.

Atlegrim, O. 1989. Exclusion of birds from bilberry stands: impact on insect larval density and damage to the bilberry. Oecologia 79:136–139.

Baltz, M. E., and S. C. Latta. 1998. Cape May Warbler (*Dendroica tigrina*). A. Poole and F. Gill (editors), The birds of North America, No. 332. The Birds of North America, Inc., Philadelphia, PA.

Blake, J. G., and B.A. Loiselle. 1992. Fruits in the diets of neotropical migrant birds in Costa Rica. Biotropica 24:200–210.

Blancher, P. 2003. Importance of Canada's boreal forest to landbirds. Canadian Boreal Initiative and Boreal Songbird Initiative, Ottawa, ON, and Seattle, WA.

Blancher, P., and J. V. Wells. 2005. The boreal forest region: North America's bird nursery. Canadian Boreal Initiative and Boreal Songbird Initiative, Ottawa, ON, and Seattle, WA.

Blancher, P. J., K. V. Rosenberg, A. O. Panjabi, B. Altman, J. Bart, C. J. Beardmore, G. S. Butcher, D. Demarest, R. Dettmers, E. H. Dunn, W. Easton, W. C. Hunter, E. E. Iñigo-Elias, D. N. Pashley, C. J. Ralph, T. D. Rich, C. M. Rustay, J. M. Ruth and T. C. Will. 2007. Guide to the Partners in Flight Population Estimates Database. Version: North American Landbird Conservation Plan 2004. Partners in Flight Technical Series No 5. http://www.rmbo.org/pif_db/laped/guide.aspx

Bryant, D., D. Nielson, and L. Tangley. 1997. The last frontier forests: ecosystems and economies on the edge. World Resources Institute, Washington, DC.

Castor, G., and I. Locker. 2000. Mapping conservation investments: An assessment of biodiversity funding in Latin America and the Caribbean. Biodiversity Support Program, Washington, DC.

Crawford, H. S., and D. T. Jennings. 1989. Predation by birds on spruce budworm *Choristoneura fumiferana*: functional, numerical, and total responses. Ecology 70:152–163.

Faaborg, J., and W. J. Arendt. 1992. Long-term declines of winter resident warblers in a Puerto Rican dry forest: which species are in trouble? Pp. 57–63 in J. M. Hagan III and D. W. Johnston (editors), Ecology and conservation of neotropical migrant landbirds. Smithsonian Institution Press, Washington, DC.

Faaborg, J., R. T. Holmes, A. D. Anders, K. L. Bildstein, K. M. Dugger, S. A. Gauthreaux, Jr., P. Heglund, K. A. Hobson, A. E. Jahn, D. H. Johnson, S. C. Latta, D. J. Levey 2nd, P. P. Marra, C. L. Merkord, E. Nol, S. I. Rothstein, T. W. Sherry, T. S. Sillett, F. R. Thompson 3rd, and N. Warnock. 2009. Conserving migratory land birds in the new world: do we know enough? Ecological Applications 20:398–418.

Holmes, R. T. 1990. Ecological and evolutionary impacts of bird predation on forest insects: an overview. Studies in Avian Biology 13:6–13.

Holmes, R. T., J. C. Schultz, and P. Nothnagle. 1979. Bird predation on forest insects: an exclosure experiment. Science 206:462–463.

Howe, H. F., and D. De Steven. 1979. Fruit production, migrant bird visitation, and seed dispersal of *Guarea glabra* in Panama. Oecologia 39:185–196.

Hutto, R. L. 1980. Winter habitat distribution of migratory land birds in western Mexico with special reference to small, foliage gleaning insectivores. Pp. 181 203 in A. Keast and E. S. Morton (editors), Migrant birds in the neotropics: ecology, behavior, distribution and conservation. Smithsonian Institution Press, Washington, DC.

Hutto, R. L. 1992. Habitat distributions of migratory landbird species in western Mexico. Pp. 211–239 in J. M. Hagan III and D. W. Johnston (editors), Ecology and conservation of neotropical migrant landbirds. Smithsonian Institution Press, Washington, DC.

Karr, J. R. 1976. On the relative abundance of migrants from the north temperate zone in tropical habitats. Wilson Bulletin 88:433–458.

Latta, S. C., C. C. Rimmer, and K. P. McFarland. 2003. Winter bird communities in four habitats along an elevational gradient on Hispaniola. Condor 105:179–197.

Lee, P. D., Aksenov, L. Laestadius, R. Nogueron, and W. Smith. 2006. Canada's large intact forest landsapes. Global Forest Watch Canada, Edmonton, AB.

Levey, D. J. 1988. Spatial and temporal variation in Costa Rican fruit and fruit-eating bird abundance. Ecological Monographs 58:251–269.

Lynch, J. F. 1992. Distribution of overwintering nearctic migrants in the Yucatan Peninsula, II: Use of native and human-modified vegetation. Pp. 178–196 in J. M. Hagan III and D. W. Johnston (editors), Ecology and conservation of Neotropical migrant landbirds. Smithsonian Institution Press, Washington, DC.

Martin, T. E. 1985. Selection of second-growth woodlands by frugivorous migrating birds in Panama: an effect of fruit size and plant density? Journal of Tropical Ecology 1:157–170.

Morton, E. S. 1971. Food and migration habits of the Eastern Kingbird in Panama. Auk 88:925–926.

Morton, E. S. 1980. Adaptations to seasonal changes by migrant landbirds in the Panama Canal Zone. Pp. 437–453 in A. Keast and E. S. Morton (editors), Migrant birds in the neotropics: ecology, behavior, distribution and conservation. Smithsonian Institution Press, Washington, DC.

Poague, K. L., R. J. Johnson, and L.J. Young. 2000. Bird use of rural and urban converted railroad rights-of-way in southeast Nebraska. Wildlife Society Bulletin 28:852–864.

Rappole, J. H. 1995. The ecology of migrant birds: a neotropical perspective. Smithsonian Institution Press, Washington, DC.

Rathcke, B. J. 2000. Hurricane causes resource and pollination limitation of fruit set in a bird-pollinated shrub. Ecology 81:1951–1958.

Reid, J. L., J. B. C. Harris, L. J. Martin, J. R. Barnett, and R. A. Zahawi. 2008. Distribution and abundance of nearctic-neotropical songbird migrants in a forest restoration site in southern Costa Rica. Journal of Tropical Ecology 24:685–688.

Reinkensmeyer, D. P., R. F. Miller, R. G. Anthony, V. E. Marr, and C. M. Duncan. 2008. Winter and early spring bird communities in grasslands, shrubsteppe, and juniper woodlands in central Oregon. Western North American Naturalist 68:25–35.

Rich, T. D., C. J. Beardmore, H. Berlanga, P. J. Blancher, M. S. W. Bradstreet, G. S. Butcher, D. W. Demarest, E .H. Dunn, W. C. Hunter, E. E. Inigo-Elias, J. A. Kennedy, A. M. Martell, A. O. Panjabi, D. N. Pashley, K. V. Rosenberg, C. M. Rustay, J. S. Wendt, and T. C. Will. 2004. Partners in Flight North American Landbird Conservation Plan. Cornell Lab of Ornithology, Ithaca, NY.

Ridgely, R. S., T. F. Allnutt, T. Brooks, D. K. McNicol, D. W. Mehlman, B. E. Young, and J. R. Zook. 2003. Digital distribution maps of the birds of the western hemisphere, version 1.0. NatureServe, Arlington, VA.

Robbins, C. S., B. A. Dowell, D. K. Dawson, J. A. Colon, R. Estrada, A. Sutton, R. Sutton, and D. Weyer. 1992. Comparison of neotropical migrant landbird populations wintering in tropical forest, isolated forest fragments, and agricultural habitats. Pp. 207–210 in J. M. Hagan III and D. W. Johnston (editors), Ecology and conservation of neotropical migrant landbirds. Smithsonian Institution Press, Washington, DC.

Slud, P. 1960. The birds of Finca "La Selva" Costa Rica: a tropical wet forest locality. Bulletin of the American Museum of Natural History 121:49–148.

Stiles, F. G. 1980. Evolutionary implications of habitat relations between permanent resident and winter resident land birds in Costa Rica. Pp. 421–435 in A. Keast and E. S. Morton (editors), Migrant birds in the neotropics: ecology, behavior, distribution and conservation. Smithsonian Institution Press, Washington, DC.

Takekawa, J. Y., and E. O. Garton. 1984. How much is an Evening Grosbeak worth? Journal of Forestry 82:426–428.

Terborgh, J. W., and J. R. Faaborg. 1980. Factors affecting the distribution and abundance of North American migrants in the eastern Caribbean region. Pp. 145–155 in A. Keast and E. S. Morton (editors), Migrant birds in the neotropics: ecology, behavior, distribution, and conservation. Smithsonian Institution Press, Washington, DC.

Torgerson, T. R., R. R. Mason, and R. W. Campbell. 1990. Predation by birds and ants on two forest insect pests in the Pacific Northwest. Studies in Avian Biology 13:14–19.

U.S. NABCI Committee. 2000. North American Bird Conservation Initiative: Bird Conservation Regions map and Bird Conservation Region descriptions. U.S. Fish and Wildlife Service, Arlington, VA.

Wallace, G. E., H. G. Alonso, M. K. Mcnicholl, D. R. Batista, R. O. Prieto, A. L. Sosa, B. S. Oria, and E. A. H. Wallace. 1996. Winter surveys of forest-dwelling neotropical migrant and resident birds in three regions of Cuba. Condor 98:745–768.

Webster, M. S., P. P. Marra, S. M. Haig, S. Bensch, and R. T. Holmes. 2002. Links between worlds: unraveling migratory connectivity. Trends in Ecology and Evolution 17:76–83.

Whelan, C. J., D. Wenny, and R. J. Marquis. 2008. Ecosystem services provided by birds. Annals of the New York Academy of Sciences: 1134:25–60.

White, D. H., C. B. Kepler, J. S. Hatfield, P. W. Sykes, Jr., and J. T. Seginak. Habitat associations of birds in the Georgia Piedmont during winter. Journal of Field Ornithology 67:159–166.

Wunderle, J. M., Jr., and R. B. Waide. 1993. Distribution of overwintering nearctic migrants in the Bahamas and Greater Antilles. Condor 95:904–933.

CHAPTER EIGHT

Important Bird Areas as Wintering Sites
for Boreal Migrants in the Tropical Andes

Ian J. Davidson, David D. Fernández and Rob Clay

Abstract. The Tropical Andes region provides important boreal winter habitat for nearctic–neotropical migrant birds, particularly birds that breed in the boreal forest of North America. Understanding of the distribution of these bird species was greatly improved through an inventory of neotropical migrants at recently identified Important Bird Areas throughout the region. Large proportions of the populations of boreal breeding species, including Olive-sided Flycatchers (*Contopus cooperi*), Eastern (*C. virens*) and Western Wood-Pewees (*C. sordidulus*), Alder Flycatchers (*Empidonax alnorum*), Blackpoll Warblers (*Dendroica striata*), Veeries (*Catharus fuscescens*), and Gray-cheeked Thrushes (*Catharus minimus*), depend on the forested slopes of the Tropical Andes. Current efforts to conserve habitat important for these birds are not considered a high priority by conservation groups in the region due to more urgent needs focused on staving off serious declines and possible extinctions of many resident and endemic critically threatened bird species. However, combined efforts to support sites of importance for both resident and migratory birds (as well as other biodiversity) may be the most effective strategy for conserving both endemic Andean species and boreal breeding migrants wintering in the Tropical Andes.

Key Words: boreal breeding migrants, conservation, Important Bird Areas, Tropical Andes.

The 1.5 billion acres of boreal forests of North America are the nesting ground for over 300 different species of birds (Blancher and Wells 2005). An estimated 90% of the boreal forest–breeding birds migrate out of the area during the boreal winter period (Rich et al. 2004). While a majority migrate to southern United States and Central America, a significant number reach the five countries of the Tropical Andes (Bolivia, Colombia, Ecuador, Peru, and Venezuela)

and almost the entire populations of some species, such as Olive-sided Flycatchers, Eastern and Western Wood-Pewees, Alder Flycatcher, Swainson's (*Catharus ustulatus*) and Gray-cheeked Thrushes, Bay-breasted (*Dendroica castanea*), Blackburnian (*D. fusca*), Canada (*Wilsonia canadensis*), and Blackpoll Warblers, and Black-billed Cuckoo (*Coccyzus erythropthalmus*), depend on the region for most of the boreal winter period (Rappole et al. 1995, NatureServe 2008).

Davidson, I. J., D. D. Fernández, and R. Clay. 2011. Important Bird Areas as wintering sites for boreal migrants in the Tropical Andes. Pp. 95–106 *in* J. V. Wells (editor). Boreal birds of North America: a hemispheric view of their conservation links and significance. Studies in Avian Biology (no. 41), University of California Press, Berkeley, CA.

Population trends indicate that of the 300 boreal breeding migrants, as many as 15% are declining (www.borealbirds.org/borealbirdspetition.shtml), two species are considered Globally Threatened, and five Near Threatened with extinction (IUCN 2010). While these declines are clearly worrying, it is uncertain whether they are a result of conditions impacting species on their breeding, migratory, or wintering habitat.

In its report, the Nature Conservancy (Roca et al. 1996) linked the importance of protected areas for neotropical migrants and, noting the declines in migrant species populations, warned that the demise of neotropical migrants raised the specter of an ever-present "silent spring" in North America. This report stressed the need for a much better understanding of how neotropical migrants are using protected areas, particularly in the Andean countries and Paraguay.

Given the large number of bird species endemic to the region and the growing threat to their habitat, most research efforts are understandably focused on this suite of bird species and not migrants. This, combined with the region's rugged geography, at times political instability, and limited availability of technical and financial resources, has resulted in little by way of standardized efforts to understand the ecology of boreal migrants in their neotropical wintering habitat. Without this knowledge, the conservation community is challenged to develop effective species plans aimed at addressing the root causes for boreal forest–breeding migrant bird population declines.

This paper summarizes recent efforts to broaden our understanding of the importance of the Tropical Andes for boreal forest–breeding migrants and presents some initial recommendations for furthering these studies and conserving critical habitat for both endemic and threatened migrants sharing the same habitat.

METHODS

In 2003, BirdLife International undertook to identify and compile information on globally Important Bird Areas (IBAs) in the Tropical Andes (BirdLife 2005). IBAs are key sites for conservation—small enough to be conserved in their entirety and often already part of a protected

areas network. They do one (or more) of three things: (1) hold significant numbers of one or more globally threatened species; (2) are one of a set of sites that together hold a suite of restricted-range species or biome-restricted species; and (3) have exceptionally large numbers of migratory or congregatory species. Once identified, BirdLife supports, promotes, and advocates for the conservation of IBAs, if these sites are not already protected.

National coordinating bodies were contracted in each country to compile baseline information on birds and sites. National and regional workshops engaged all key stakeholders in the review and validation of baseline studies as well as the identification of IBAs. Information collected and collated in each of the five countries was entered into a database managed by BirdLife, which allows for the management and analysis of this baseline data. At each of the national workshops, IBA boundaries were delineated and mapped at a scale ranging from of 1:50,000 to 1:250,000, depending on size. This information was digitized and transposed onto GoogleEarth© for viewing purposes.

IBAs then formed the basis for the study. Information on neotropical (i.e., nearctic-breeding) migrants found in IBAs was compiled as part of the national IBA workshop process. Where information was lacking or considered questionable, additional efforts were made to gather information from various sources, including the following: primary or gray literature, regional bird guides, field trip reports, and expert information. In all situations, this information was carefully reviewed, and where there were questions about validity, this information was not included.

The total number of boreal forest–breeding migrants recorded in the Tropical Andes was based on a list of migrants compiled from the IBA species lists. Where possible, abundance data was gathered, even if anecdotal, since very little if any data on this characteristic of boreal forest breeding migrants was readily available.

A website was created to present data on (1) species (including ecology, range distribution, threat and conservation status, abundance, and photographs), (2) IBAs (including site description, conservation and threat status, and maps), and (3) relevant literature and web links.

RESULTS

Boreal-Breeding Migrant Birds in the Tropical Andes

Of the 341 neotropical migrants recognized by the USFWS (2010) in the Americas, 159 occur in the Tropical Andes. This includes 123 migrants that partially or wholly breed in the boreal forest region (Appendix 8.1). About 28 of the 123 boreal forest–breeding migrants have more than 50 percent of their winter distribution in the Tropical Andes (either resident during the boreal winter period or transient to more southern habitats). Colombia and Venezuela host the largest number of boreal forest breeding migrants, with the number of species declining rapidly moving southward (Table 8.1).

Several species—including Olive-sided Flycatchers; Eastern and Western Wood-Pewees; Alder Flycatchers; Swainson's and Gray-cheeked Thrushes; Connecticut, Mourning (*Oporornis philadelphia*), Canada, Blackburnian, Chestnut-sided (*Dendroica pensylvanica*), Bay-breasted, Tennessee (*Vermivora peregrina*), and Blackpoll Warblers; Northern Waterthrushes (*Seiurus noveboracensis*); Black-billed Cuckoos; Rose-breasted Grosbeaks (*Pheucticus ludovicianus*); Scarlet Tanagers (*Piranga olivacea*), and Common Nighthawks (*Chordeiles minor*)—are relatively common in Colombia, Venezuela, and Ecuador, especially in secondary forests between 500 and 2,000 meters. The eastern Andean foothills boast the greatest number of species, but species numbers decline moving southward toward Bolivia. The Amazon basin was surprisingly depauperate of boreal forest–breeding migrants except for the common Blackpoll Warbler, which appears to migrate through the Andes to lowland Amazon tropical forests (below 500–1,000 m) of the foothills adjacent to the Andes as well as around the Tepuis of southern Venezuela (Fig. 8.1).

Boreal forest–breeding shorebirds, including Greater (*Tringa melanoleuca*) and Lesser Yellowlegs (*Tringa flavipes*), Whimbrel (*Numenius phaeopus*), Semipalmated Plover (*Charadrius semipalmatus*), Solitary Sandpiper (*Tringa solitaria*), Spotted Sandpiper (*Actitis macularius*), Least Sandpiper (*Calidris minutilla*), and Short-billed Dowitcher (*Limnodromus griseus*), move along the Pacific coast, staging at times at productive coastal lagoons. Particularly large numbers of Semipalmated Plovers (flocks of

TABLE 8.1
Number of boreal forest breeding migrants in Tropical Andean countries

Tropical Andean Country	No. of Boreal Forest Breeding Migrants
Venezuela	115
Colombia	123
Ecuador	105
Peru	59
Bolivia	45

5,000 or more) have been observed in coastal mangrove estuaries in northern Ecuador. The same suite of shorebirds occurs in smaller flocks on high Andean lagoons and in wet Páramo grasslands. The Upland Sandpiper (*Bartramia longicauda*) appears to move along both sides of the Andes, with observations of small flocks in Ecuador apparently attempting to cross the Andes near Loja, where several hundreds have been observed to drown in the higher Andean lagoons (pers. obs.).

Common (*Sterna hirundo*) and Black Terns (*Chlidonias niger*), as well as smaller numbers of Caspian Terns (*Sterna caspia*), make their way southward along the coast to Ecuador and central Peru and eastward along the Caribbean coast of Venezuela. Franklin's Gulls (*Larus pipixcan*) are also common along the entire Pacific coast of South America, often congregating around ports where refuse from fishing boats and factories is dumped. Broad-winged Hawks (*Buteo platypterus*) seem to be common through the entire region and can been seen in small flocks riding thermals along the eastern Andes.

While some species, like the Olive-sided Flycatcher, breed almost entirely in the boreal forest, other species including Common and Black Terns breed in a variety of other habitats across Canada and the United States. Since banding data or other information that may help establish geographic origin (e.g., subspecific identity, stable isotope analyses) are sparse, it is difficult to ascertain if it is actually boreal forest–breeding birds that winter in the Tropical Andean region. Recent intensive banding efforts by some groups in Colombia and Venezuela may help to provide recapture data that will further pinpoint the origin of many species

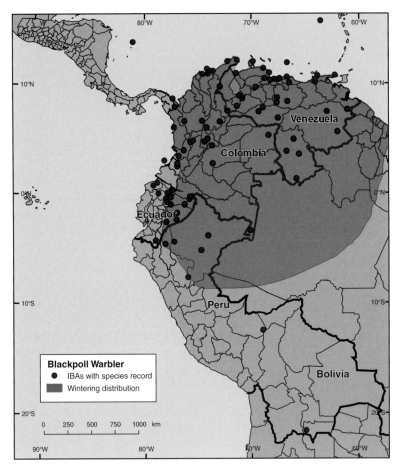

Figure 8.1. Winter range of Blackpoll Warbler (opaque gray area), showing IBAs (dark gray dots) where the species has been recorded.

and confirm whether they have their origin in the boreal forest or not.

Species Distributions in the Tropical Andes

Using NatureServe (2008) maps of species distribution, this study found range extensions for nearly 50% of the boreal forest–breeding migrants. Some of these were considerable range extensions, as is the case of the Blackpoll Warbler (Fig. 8.1), which was recorded almost 1,000 kilometers further south than previously mapped. In other cases, the study helped refine species distribution—for example, the Blackpoll Warbler, which appears to prefer the tropical forests adjacent to the eastern Andes, whereas NatureServe maps have the Blackpoll Warbler distributed throughout the western Amazon basin (Fig. 8.1).

Habitat Status in the Tropical Andes

Along the eastern and western flanks as well as the central valleys of the Andes in Venezuela, Colombia, and Ecuador, forest habitat has largely been replaced by agricultural crops (e.g., beans, corn, coffee, vegetables) and ranching. Only Colombia's Choco forests along the southwest coast and the dry forests of the northwest coast of Colombia and Venezuela remain relatively intact. Information gathered from IBAs in this region indicates that agricultural expansion is the main threat to important areas for breeding migrants. Large extensions of subtropical and tropical forest remain along the eastern Andes from central Ecuador southward to the Bolivian border with Peru. However, concentrations of most boreal-breeding migrants (particularly passerine species) decrease rapidly from central Ecuador southward.

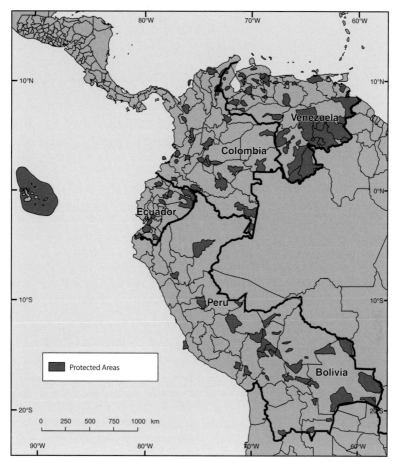

Figure 8.2. Map of protected areas in the Tropical Andes region.

A majority of the key wetlands along the Caribbean have been converted to other land uses, while adjacent dry forests have been largely cleared of primary forests. Much of the coastal wetlands of Ecuador have also been significantly altered for shrimp and rice production. Those that have not been altered are severely contaminated with untreated sewage from nearby towns and agricultural chemicals from farmlands, as well as industrial waste from large urban areas in the central cordillera.

DISCUSSION

Of the more than 3,000 species of birds occurring in the Tropical Andes region, a small percentage of these are migrants and an even smaller portion depend on the boreal forest for the breeding period. Nonetheless, the Tropical Andes provides wintering habitat for nearly the entire populations of many boreal-breeding migrants, including the Olive-sided Flycatcher, Eastern and Western Wood-Pewees, Alder Flycatcher, Broad-winged Hawk, Semipalmated Plover, Blackpoll Warbler, Veery, Gray-cheeked and Swainson's Thrushes, and Chimney Swift. To ensure their conservation, it is vital that their habitat preferences as well as drivers impacting the integrity of these habitats are understood. To date, there is a fair amount of anecdotal information on species habits but only very coarse data on their distribution and habitat preferences. Efforts to systematically compile and verify existing information from the Tropical Andes is considered a priority to better understand the particular needs of these and other boreal migrants.

Of particular concern for U.S. and Canadian conservationists is the plight of several species listed on the IUCN Red List (IUCN 2010) (e.g., Olive-sided Flycatcher) and/or categorized as Birds of Conservation Concern (USFWS 2002) (Appendix 8.1).

Protected areas systems support a variety of habitat types in each of the five Tropical Andean countries (Fig. 8.2). Government support for these protected areas, while increasing over recent years, still is wholly insufficient, and as a result many protected areas are at risk. Fortunately, bilateral and multilateral agencies, together with governments in the region, recognize the importance of protected areas as a source of employment (e.g., through tourism) and for their ecological services (e.g., provision of water) and are increasingly predisposed to making funds available for integrating human and biological needs for these areas.

Through a project supported by the U.S. Fish and Wildlife Service, BirdLife International has incorporated data on neotropical migrants into a global database on IBAs. These data reveals that as much as 45% of IBAs are found outside of protected areas systems and, therefore, many of these globally threatened sites have no protection at all. Unprotected sites include many where congregatory boreal forest–breeding migrants like shorebirds and waterbirds stage.

And while individual IBAs do not help protect significant numbers of boreal forest breeding migrants, collectively, the conservation of a complex of IBAs could (e.g., a series of IBAs along the eastern flanks of the Andes between Colombia and Ecuador could provide sufficient habitat for a significant proportion of Blackburnian and Canada Warbler populations). The conservation of individual IBAs is more likely to help conserve congregatory boreal forest–breeding migrants (e.g., both species of Yellowlegs, Semipalmated Sandpiper, Whimbrel, and Franklin's Gull). It is also conceivable that select sites along the Colombian and Venezuelan Caribbean coast (e.g., Henri Pittier National Park) may be extremely important as migrants reach land after their perilous journey over the Caribbean from North America. However, aside from years of banding work at Henri Pittier National Park, little is known about the importance of other forested sites along the coast.

One of the challenges of conserving boreal forest–breeding migrants outside of their breeding grounds is that they are not considered a priority. With nearly 10% of the resident neotropical birds classified as globally threatened, the priority is focused on preventing extinctions. It is estimated that in the Tropical Andes we risk losing as many as 15–20 Critically Threatened bird species over the next 5–10 yrs. Not surprisingly, then, the conservation community is focusing on this suite of highly imperiled birds.

There does exist an opportunity to protect boreal forest–breeding migrants by focusing on sites that support resident species that are also important for migratory species. An example of this is the conservation of a subtropical forest at 2,000 m in eastern Ecuador called the Huacamayas IBA, where efforts are under way with funds for neotropical bird conservation to conserve habitat for resident and endemic species like the Peruvian (*Grallaricula peruviana*) and Giant Antpittas (*Grallaria gigantea*), Bicolored Antvireo (*Dysithamnus occidentalis*) and a (sub)species of owl (*Strix* sp.) still to be described, while at the same time helping to ensure suitable forest habitat shared also by Blackburnian, Mourning, and Canada Warblers, and Swainson's Thrush.

For the first time, IBAs provide a means to focus conservation efforts at a discrete set of sites for the conservation of boreal forest–breeding migrants in addition to threatened and endemic resident species. Because of the dispersed nature of most boreal forest–breeding birds in the Tropical Andes, it is unlikely that the conservation of an individual IBA will impact the status of any boreal forest–breeding birds. However, the conservation of a suite of IBAs in a particular region or forest habitat type may have a positive impact for migratory species, especially along the largely deforested Tropical Andean slopes of Venezuela, Colombia, and Ecuador. In addition, species of boreal forest–breeding birds that concentrate at IBAs in large numbers (e.g., Franklin's Gulls) will benefit from ongoing and new efforts to conserve these IBAs. Overall, if a large percentage of IBAs in any given Tropical Andean country were conserved, it could secure significant tracts of key wintering and passage habitat for many boreal forest–breeding birds.

Research efforts to further understand the specific habitat requirements of boreal forest–breeding migrant birds are urgently required. Current distribution maps provide a coarse understanding of species distribution. Initial work on other species of neotropical migrants (e.g., the Cerulean Warbler, *Dendroica cerulea*) demonstrates that habitat use not only varies as the species migrate into the Andes, but is also very specific to certain altitudes and microhabitats during the middle of the boreal winter period. Systematic surveys undertaken by students, birdwatchers, birding guides and even local communities would help to better gather data on species presence and abundance. Current banding efforts for neotropical migrants are under way in Colombia and Venezuela. The expertise developed at these sites could serve as the basis for training other Andean-based biologists in appropriate banding techniques. Intensive banding efforts undertaken at key IBAs in the Tropical Andes would also help to determine the origin of many boreal forest breeding migrants.

Resources are critical to facilitate both research and conservation work. The Neotropical Migratory Bird Conservation Act provides vital resources that can help lever additional in-country resources for boreal forest–breeding and resident bird conservation. IBAs provide a focus for targeting scarce resources for the conservation of migrant and resident species as well as other species in this biodiversity hotspot.

ACKNOWLEDGMENTS

The authors would like to thank the following people and institutions for their financial and technical support: D. Ryan, USFWS; L. Suarez and J.V. Rodriguez, Conservation International; T. Santander, Aves & Conservacion; B. Hennessey, Asociacion Armonia; M. Lentino, SCAV, IAvH, Lima Museum of Natural History; K. Boyla, C. Devenish, P. Salaman, and R. Williams.

LITERATURE CITED

BirdLife International. 2005. Important Bird Areas of the Tropical Andes. BirdLife International. Quito, Ecuador. <http://www.birdlife.org/action/science/sites/neotrops/andes/index.html?language=en>.

Blancher, P., and J. Wells 2005. The boreal forest region: North America's bird nursery. Boreal Songbird Initiative, Ottawa, ON.

IUCN Red List. 2008. <http://cms.iucn.org/about/work/programmes/species/red_list>.

Kushlan, J. A., M. J. Steinkamp, K. C. Parsons, J. Capp, M. Acosta Cruz, M. Coulter, I. Davidson, L. Dickson, N. Edelson, R. Elliot, R. Michael Erwin, S. Hatch, S. Kress, R. Mildo, S. Miller, K. Mills, R. Paul, R. Phillips, J. E. Salva, B. Sydeman, J. Trapp, J. Wheeler, and K. Wohl. 2002. Waterbird conservation for the Americas: the North American waterbird conservation plan, version 1. Waterbird Conservation for the Americas, Washington, DC.

NatureServe. 2008. <http://www.natureserve.org>.

North American Waterfowl Management Plan Committee. 2004. North American Waterfowl Management Plan 2004: implementation framework: strengthening the biological foundation. Canadian Wildlife Service, U.S. Fish and Wildlife Service, Secretaria de Medio Ambiente y Recursos Naturales.

Rappole, J. H., E. S. Morton, and T. E. Lovejoy III. 1995. Nearctic avian migrants in the neotropics: maps, appendices, and bibliography. 2nd ed. Smithsonian Institution, Front Royal, VA.

Rich, T. D., C. J. Beardmore, H. Berlanga, P. J. Blancher, M. S. W. Bradstreet, G. S. Butcher, D. W. Demarest, E. H. Dunn, W. C. Hunter, E. E. Inigo-Elais, J. A. Kennedy, A. M. Martell, A. O. Panjabi, D. N. Pashley, K. V. Rosenberg, C. M. Rustay, J. S. Wendt, and T. C. Will. 2004. Partners in Flight North American landbird conservation plan. Cornell Lab of Ornithology. Ithaca, NY. <http://www.partnersinflight.org/cont_plan/> (March 2005).

Roca, R,, L. Adkins, M. C. Wurschy, and K. L. Skerl. 1996. Wings from afar: an ecoregional approach to conservation of neotropical migratory birds in South America. The Nature Conservancy, Arlington, VA.

U.S. Fish and Wildlife Service. 2002. Birds of conservation concern 2002. Division of Migratory Bird Management, Arlington, VA. <http://migratorybirds.fws.gov/reports/bcc2002.pdf>.

U.S. Shorebird Conservation Plan. 2004. High priority shorebirds 2004. Unpublished report. U. S. Fish and Wildlife Service, Arlington, VA.

Neotropical migrants that occur in the Tropical Andes

Species	Birds of Conservation Concern 2008	IUCN Red List 2010	Status in N.A. Conservation Plans[a]
American Wigeon (*Anas americana*)		LC	Mod. high
Blue-winged Teal (*Anas discors*)		LC	Mod. high
Northern Shoveler (*Anas clypeata*)		LC	Moderate
Northern Pintail (*Anas acuta*)		LC	High
Green-winged Teal (*Anas crecca*)		LC	Moderate
Ring-necked Duck (*Aythya collaris*)		LC	Moderate
Lesser Scaup (*Aythya affinis*)		LC	High
Great Blue Heron (*Ardea herodias*)		LC	Not at risk
Green Heron (*Butorides virescens*)		LC	Low
Turkey Vulture (*Cathartes aura*)		LC	
Merlin (*Falco columbarius*)		LC	
Peregrine Falcon (*Falco peregrinus*)	x	LC	
Osprey (*Pandion haliaetus*)		LC	
Northern Harrier (*Circus cyaneus*)		LC	
Broad-winged Hawk (*Buteo platypterus*)		LC	
Swainson's Hawk (*Buteo swainsoni*)	x	LC	14 (PIF)
Red-tailed Hawk (*Buteo jamaicensis*)		LC	
Sora (*Porzana carolina*)		LC	
American Golden-Plover (*Pluvialis dominica*)		LC	High concern
Black-bellied Plover (*Pluvialis squatarola*)		LC	
Semipalmated Plover (*Charadrius semipalmatus*)		LC	
Killdeer (*Charadrius vociferus*)		LC	
Piping Plover (*Charadrius melodus*)		NT	Highly imperiled
Wilson's Snipe (*Gallinago gallinago*)		LC	
Hudsonian Godwit (*Limosa haemastica*)	x	LC	High concern
Marbled Godwit (*Limosa fedoa*)	x	LC	High concern
Whimbrel (*Numenius phaeopus*)	x	LC	High concern[b]
Upland Sandpiper (*Bartramia longicauda*)	x	LC	High concern
Greater Yellowlegs (*Tringa melanoleuca*)		LC	
Lesser Yellowlegs (*Tringa flavipes*)	x	LC	
Solitary Sandpiper (*Tringa solitaria*)	x	LC	High concern
Spotted Sandpiper (*Actitis macularius*)		LC	
Wandering Tattler (*Heteroscelus incanus*)		LC	

APPENDIX 8.1 (*continued*)

Species	Birds of Conservation Concern 2008	UICN Red List 2010	Status in N.A. Conservation Plans[a]
Willet (*Catoptrophorus semipalmatus*)		LC	
Ruddy Turnstone (*Arenaria interpres*)		LC	High concern[b]
Short-billed Dowitcher (*Limnodromus griseus*)	x	LC	High concern
Long-billed Dowitcher (*Limnodromus scolopaceus*)		LC	
Surfbird (*Aphriza virgata*)		LC	High concern
Red Knot (*Calidris canutus*)	x	LC	Highly imperiled[b,c]
Sanderling (*Calidris alba*)		LC	High concern[b]
Semipalmated Sandpiper (*Calidris pusilla*)	x	LC	
Western Sandpiper (*Calidris mauri*)		LC	High concern
Least Sandpiper (*Calidris minutilla*)		LC	
White-rumped Sandpiper (*Calidris fuscicollis*)		LC	
Baird's Sandpiper (*Calidris bairdii*)		LC	
Pectoral Sandpiper (*Calidris melanotos*)		LC	
Dunlin (*Calidris alpina*)	x	LC	High concern[b]
Stilt Sandpiper (*Micropalama himantopus*)		LC	
Buff-breasted Sandpiper (*Tryngites subruficollis*)	x	NT	Highly imperiled
Wilson's Phalarope (*Steganopus tricolor*)		LC	High concern
Red-necked Phalarope (*Phalaropus lobatus*)		LC	
Red Phalarope (*Phalaropus fulicarius*)		LC	
Ring-billed Gull (*Larus delawarensis*)		LC	Not at risk
Great Black-backed Gull (*Larus marinus*)		LC	Not at risk
Herring Gull (*Larus argentatus*)		LC	Low
Bonaparte's Gull (*Larus philadelphia*)		LC	Moderate
Franklin's Gull (*Larus pipixcan*)		LC	Moderate
Sabine's Gull (*Xema sabini*)		LC	Low
Caspian Tern (*Sterna caspia*)		LC	Low
Common Tern (*Sterna hirundo*)		LC	Low
Arctic Tern (*Sterna paradisaea*)		LC	High
Black Tern (*Chlidonias niger*)		LC	Moderate
Parasitic Jaeger (*Stercorarius parasiticus*)		LC	Low
Long-tailed Jaeger (*Stercorarius longicaudus*)		LC	Low
Black-billed Cuckoo (*Coccyzus erythropthalmus*)		LC	
Common Nighthawk (*Chordeiles minor*)		LC	
Chimney Swift (*Chaetura pelagica*)		NT	
Belted Kingfisher (*Ceryle alcyon*)		LC	
Yellow-bellied Sapsucker (*Sphyrapicus varius*)		LC	
Olive-sided Flycatcher (*Contopus cooperi*)	x	NT	14 (PIF)
Western Wood-Pewee (*Contopus sordidulus*)		LC	

APPENDIX 8.1 (*continued*)

Species	Birds of Conservation Concern 2008	UICN Red List 2010	Status in N.A. Conservation Plans[a]
Eastern Wood-Pewee (*Contopus virens*)		LC	
Alder Flycatcher (*Empidonax alnorum*)		LC	
Great Crested Flycatcher (*Myiarchus crinitus*)		LC	
Eastern Kingbird (*Tyrannus tyrannus*)		LC	
Philadelphia Vireo (*Vireo philadelphicus*)		LC	
Red-eyed Vireo (*Vireo olivaceus*)		LC	
Cedar Waxwing (*Bombycilla cedrorum*)		LC	
Tree Swallow (*Tachycineta bicolor*)		LC	
Violet-green Swallow (*Tachycineta thalassina*)		LC	
Purple Martin (*Progne subis*)		LC	
Bank Swallow (*Riparia riparia*)		LC	
Barn Swallow (*Hirundo rustica*)		LC	
Cliff Swallow (*Petrochelidon pyrrhonota*)		LC	
Gray Catbird (*Dumetella carolinensis*)		LC	
Veery (*Catharus fuscescens*)		LC	
Gray-cheeked Thrush (*Catharus minimus*)		LC	
Swainson's Thrush (*Catharus ustulatus*)		LC	
Wood Thrush (*Hylocichla mustelina*)	x	LC	14 (PIF)
Tennessee Warbler (*Vermivora peregrina*)		LC	
Northern Parula (*Parula americana*)		LC	
Yellow Warbler (*Dendroica petechia*)		LC	
Chestnut-sided Warbler (*Dendroica pensylvanica*)		LC	
Magnolia Warbler (*Dendroica magnolia*)		LC	
Cape May Warbler (*Dendroica tigrina*)		LC	
Black-throated Blue Warbler (*Dendroica caerulescens*)		LC	
Yellow-rumped Warbler (*Dendroica coronata*)		LC	
Townsend's Warbler (*Dendroica townsendi*)		LC	
Black-throated Green Warbler (*Dendroica virens*)		LC	
Blackburnian Warbler (*Dendroica fusca*)		LC	
Pine Warbler (*Dendroica pinus*)		LC	
Palm Warbler (*Dendroica palmarum*)		LC	
Bay-breasted Warbler (*Dendroica castanea*)	x	LC	14 (PIF)
Blackpoll Warbler (*Dendroica striata*)		LC	
Black-and-white Warbler (*Mniotilta varia*)		LC	
American Redstart (*Setophaga ruticilla*)		LC	
Prothonotary Warbler (*Protonotaria citrea*)	x	LC	15 (PIF)
Ovenbird (*Seiurus aurocapilla*)		LC	

APPENDIX 8.1 (*continued*)

Species	Birds of Conservation Concern 2008	UICN Red List 2010	Status in N.A. Conservation Plans[a]
Northern Waterthrush (*Seiurus noveboracensis*)		LC	
Louisiana Waterthrush (*Seiurus motacilla*)		LC	
Kentucky Warbler (*Oporornis formosus*)	x	LC	14 (PIF)
Connecticut Warbler (*Oporornis agilis*)		LC	
Mourning Warbler (*Oporornis philadelphia*)		LC	
Common Yellowthroat (*Geothlypis trichas*)		LC	
Wilson's Warbler (*Wilsonia pusilla*)		LC	
Canada Warbler (*Wilsonia canadensis*)	x	LC	14 (PIF)
Baltimore Oriole (*Icterus galbula*)		LC	
Bobolink (*Dolichonyx oryzivorus*)		LC	
Indigo Bunting (*Passerina cyanea*)		LC	
Scarlet Tanager (*Piranga olivacea*)		**LC**	
Rose-breasted Grosbeak (*Pheucticus ludovicianus*)		**LC**	

NOTE: Species marked in bold have 50% or more of their wintering distribution within the five countries of the Tropical Andes.

[a] Conservation plans consulted:

[b] North American populations.

[c] North American populations other than Canadian Arctic–Atlantic coast population considered of high concern; PIF watchlist species of continental concern scoring 13+.

CHAPTER NINE

Understanding Declines in Rusty Blackbirds

Russell Greenberg, Dean W. Demarest, Steven M. Matsuoka,
Claudia Mettke-Hofmann, David Evers, Paul B. Hamel,
Jason Luscier, Luke L. Powell, David Shaw, Michael L. Avery,
Keith A. Hobson, Peter J. Blancher, and Daniel K. Niven

An enormously abundant migrant. . . . The thousands of Grackles have been increased to tens of thousands. They blacken the fields and cloud the air. The bare trees on which they alight are foliated by them. Their incessant jingling songs drown the music of the Meadow Larks and produce, dreamy, far-away-effect, as of myriads of distant sleigh bells.

E. E. THOMPSON (1891), *Birds of Manitoba*

Abstract. The Rusty Blackbird (*Euphagus carolinus*), a formerly common breeding species of boreal wetlands, has exhibited the most marked decline of any North American landbird. North American Breeding Bird Survey (BBS) trends in abundance are estimated to be −12.5%/yr over the last 40 years, which is tantamount to a >95% cumulative decline. Trends in abundance calculated from Christmas Bird Counts (CBC) for a similar period indicate a range-wide decline of −5.6%/yr. Qualitative analyses of ornithological accounts suggest the species has been declining for over a century. Several studies document range retraction in the southern boreal forest, whereas limited data suggest that abundance may be more stable in more northerly areas. The major hypotheses for the decline include degradation of boreal habitats from logging and agricultural development, mercury contamination, and wetland desiccation resulting from global warming. Other likely reasons for decline include loss or degradation of wooded wetlands of the southeastern U.S and mortality associated with abatement efforts targeting nuisance blackbirds. In addition, the patchy breeding distribution of this species may inhibit population consolidation, causing local populations to crash when reduced to low levels. Progress in understanding the causes and mechanisms for observed declines has remained limited until recently. Here we present initial attempts to understand the habitat requirements of Rusty Blackbirds and offer specific predictions associated with each of the hypotheses for decline as a way of guiding future research.

Key Words: contaminants, *Euphagus carolinus*, habitat use, limiting factors, population decline, population movements, Rusty Blackbird.

Greenberg, R., D. W. Demarest, S. M. Matsuoka, C. Mettke-Hofmann, D. Evers, P. B. Hamel, J. Luscier, L. L. Powell, D. Shaw, M. L. Avery, K. A. Hobson, P. J. Blancher, and D. K. Niven. 2011. Understanding declines in Rusty Blackbirds. Pp. 107–126 *in* J. V. Wells (editor). Boreal birds of North America: a hemispheric view of their conservation links and significance. Studies in Avian Biology (no. 41), University of California Press, Berkeley, CA.

The boreal zone provides the most extensive forested habitat for high-latitude birds. Because large parts of the region are inaccessible by road, even large changes in the status of a boreal forest species may go unnoticed, or if detected, remain challenging to investigate and understand. The Rusty Blackbird (*Euphagus carolinus*) is a widespread boreal breeding songbird that has undergone a precipitous decline, as evidenced by data collected through breeding and wintering surveys from across its North American range (Greenberg and Droege 1999, Niven et al. 2004, Sauer et al. 2005). Unlike many migratory species that breed in remote boreal habitats, the Rusty Blackbird winters entirely in temperate North America, providing an opportunity to monitor the status of the entire population and evaluate how it may be responding to threats occurring throughout the year. However, the inconspicuous behavior of Rusty Blackbirds, coupled with their use of relatively inaccessible habitats during winter (e.g., forested wetlands and swamps), complicate efforts to assess status on the less remote wintering grounds. Thus, despite the fact that the decline of Rusty Blackbirds has spanned several decades and has been widely recognized for over fifteen years (Avery 1995, Link and Sauer 1996, Greenberg and Droege 1999), only very recently has research attempted to understand and address the causes. In this paper, we summarize what we currently know or strongly suspect about the basic ecology and conservation of Rusty Blackbirds, describe ongoing efforts to fill critical information gaps and present a research strategy for future work on this species. We present this paper because of the intrinsic importance of understanding and addressing such a concerning decline in a formerly common and widespread bird, and to provide insights to approaches that might be applicable to other boreal species that present some of the same research challenges.

GENERAL DISTRIBUTION AND ECOLOGY

An estimated 80–90% of all Rusty Blackbirds breed across the boreal forest region of North America (Blancher and Wells 2005), from Alaska to Newfoundland and south into the Maritime Provinces, Adirondack Mountains, and the coastal rainforest zone of southeastern Alaska (Kessel and Gibson 1978, Godfrey 1986, Avery

1995). Breeding is closely tied to forested or tall shrubby wetlands and riparian zones (Erskine 1977, Avery 1995), with birds remaining largely absent from adjacent upland interior forests and shrublands (Whitaker and Montevecchi 1997, 1999). Rusty Blackbirds winter almost entirely in temperate North America, where the core wintering area is located within the southeastern United States (Avery 1995). The species winters primarily in shallowly flooded wooded wetlands of the Mississippi Alluvial Valley and South Atlantic Coastal Plain. Current population estimates developed using data and extrapolations from North American Breeding Bird Survey (BBS), the Canadian Breeding Bird Census Database, and other sources range from 158,000 to 2 million individuals (Rich et al. 2004, Savignac 2006), and are strongly influenced by the validity of a few key assumptions that have not been rigorously evaluated (Rosenberg and Blancher 2005, Thogmartin et al. 2006).

EVIDENCE FOR THE DECLINE

Analyses of long-term data sets including the BBS (Sauer et al. 2005) and Christmas Bird Count (CBC) (Niven et al. 2004) have documented consistent and significant declines in Rusty Blackbirds over the past 40 years. Additional careful review of historical accounts (Greenberg and Droege 1999) suggests that Rusty Blackbirds had already gone from conspicuously abundant to uncommon in many areas even before these modern survey efforts began tracking them. Collectively, these observations and data describe alarming and sustained population declines, range retractions, and local extirpations from across the range.

North American Breeding Bird Survey

BBS data currently provide the only standardized long-term assessment of large-scale breeding season abundance of the Rusty Blackbird. For the period 1966–2005, these data indicate a survey-wide population decline that averages approximately $-12.5\%/\text{yr}$ ($\text{CI}_{95\%} \pm 6.3\%/\text{yr}$, $P < 0.01$; Table 9.1, Fig. 9.1) (Sauer et al. 2005). This trend corresponds to a loss of >95% of the population since 1966, and represents one of the largest population declines documented by the BBS (Link and Sauer 1996, Sauer et al. 2005).

TABLE 9.1
North American Breeding Bird Survey annual trend estimates (%/yr) for Rusty Blackbird, 1966–2005
(Sauer et al. 2005)

Region	Trend[a]	P[b]	n[c]	Mean Birds/Route
Alaska[d]	−5.3	0.04	27	0.84
Yukon Territory[d]	−12.8	0.05	7	0.32
British Columbia	−33.0	0.21	7	0.05
Ontario	−14.9	0.01	11	0.24
Quebec	−9.8	0.01	15	0.46
Newfoundland[d]	−7.7	0.20	15	2.03
New Brunswick	−8.9	0.02	17	0.28
Nova Scotia	−3.8	0.29	20	0.53
Maine	28.0	0.31	9	0.11
New Hampshire	−0.2	0.90	6	0.11
New York	2.5	0.70	7	0.07
Surveywide[e]	−12.5	<0.01	97	0.26

[a] Data for population trends are considered deficient in either sample sizes ($n < 14$ routes) or the mean number of detected birds per route (<1.0 birds/route; Sauer et al. 2005).

[b] Probability that the estimated trend differs from 0%/yr.

[c] Number of routes included in the analyses.

[d] Trends from Alaska, Yukon Territory, and Newfoundland are based on data from 1980–2005.

[e] Surveywide trend analysis excludes data from Alaska, Yukon Territory, Newfoundland, and northern portions of some provinces.

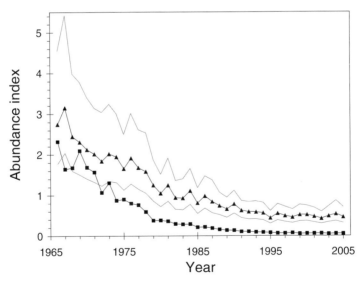

Figure 9.1. Trends in the abundance index of Rusty Blackbirds from 1966 to 2005 as estimated from the Christmas Bird Count (▲, with 97.5% credible interval in gray) and North American Breeding Bird Survey (■; Sauer et al. 2005). Christmas Bird Count data were from Bird Conservation Regions where the species was recorded on ≥4 count circles (Niven et al. 2004).

Although more variable and less precise than the survey-wide trend estimate, regional BBS analyses are also demonstrative of widespread declines (Table 9.1).

The magnitude, significance, and apparent long-term consistency of the survey-wide BBS trend for Rusty Blackbird provide compelling but somewhat limited evidence for a sharp range-wide downturn in population size. First, Rusty Blackbirds are detected on relatively few BBS routes (approximately 150), many of which have been surveyed sporadically and have frequent observer turnover. Combined with the low average number of detections per route, these factors contribute to a relatively wide confidence interval (i.e., poor precision) for the trend estimate. Second, survey coverage is limited to <30% of the breeding range and is concentrated in the southern portion of the boreal forest. Thus, the possibility exists that survey-wide BBS data are not representative of trends in more northerly breeding areas. For example, the survey-wide trend estimates exclude data from survey routes in Alaska, Newfoundland, Yukon Territory, and northern portions of some provinces because few routes in these areas encompass the long-term period of analysis (Bystrak 1981, Sauer et al. 2005). Although regional trend analyses for these northern areas do indicate declines of magnitude similar to the survey-wide estimate (Table 9.1), the precision of regional trend estimates is thought to be considerably poorer than the survey-wide estimate (Sauer et al. 2005).

Christmas Bird Count

The CBC is an invaluable source of data on the status of Rusty Blackbirds because count circles are distributed across the entire winter range of the species. Thus, we estimated winter population trend from CBC data for the period 1966–2005 following the methods of Niven et al. (2004). The range-wide trend estimate for Rusty Blackbird is −4.5%/yr (95% credible interval ± 1.2%/yr; Fig. 9.1). Though of smaller magnitude than the BBS trend estimate for the same time period (−12.5%/yr), this decline is significant and tantamount to a total decline of approximately 85% for the 40-year period. This estimate is based on data from 1,611 count circles. This sample size is an order of magnitude greater than the number of BBS routes with Rusty Blackbird detections

(i.e., 150) and may contribute to the smaller confidence interval on the CBC trend estimate. CBC data indicate a decline in all of the Bird Conservation Regions (BCR) where Rusty Blackbirds were detected. In particular, a strong and consistent (3.7–5.1%/yr) decline was estimated for the four BCRs with the highest relative abundance (Central Hardwoods, West Gulf Coastal Plain, Southeastern Coastal Plain, and Mississippi Alluvial Valley).

Two patterns are evident in the CBC data and may provide insight into the Rusty Blackbird decline (Fig. 9.1). First, the rate of decline has diminished in recent decades. The estimated annual decline over the ten-year period from 1994 to 2003 is only −2.1%/yr (Savignac 2006), whereas a period of marked decline occurred in the early 1970s. This observation suggests that factors contributing to declines in Rusty Blackbirds may have been particularly profound during or immediately preceding the 1970s. Second, the relative annual variation in counts was much greater prior to the late 1970s, and fluctuations around the estimated trend line have dampened as the population has continued to drop. Though variability might inherently diminish as abundance approaches zero, this pattern is worth further exploration as it may indicate that natural population cycling is no longer occurring—or that the ability of existing programs to detect significant population changes with such low numbers is rapidly approaching effort-related limits.

In comparing the two surveys, CBC data are less constrained by limited geographic coverage and small sample size than are those from BBS; thus the trends estimated from CBC may have greater external validity. On the other hand, the sampling effort for the BBS is far more carefully controlled, meaning that BBS data may have greater internal validity. Further, because the two surveys track populations at different points in the annual cycle, valid biological explanations may account for the differences in trend estimates. For example, the less severe decline suggested by the CBC trend may reflect tempering of the trend estimate by the annual production of young available for counting during early winter, whereas the BBS trend is based largely on overwinter survival and the recruitment of adults into the breeding population. Specifically, if fecundity or winter survivorship of Rusty Blackbirds increases as populations or densities decrease, then higher reproductive

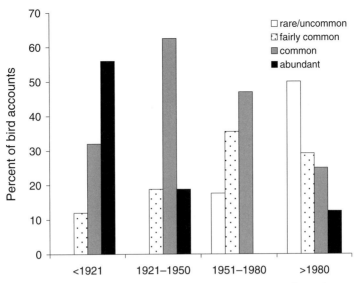

Figure 9.2. Percent of regional or state ornithological accounts ($n = 86$) listing the Rusty Blackbird in four different abundance classes and time periods.

output and early winter survivorship resulting from breeding population declines could potentially bias early winter trend estimates upward, effectively dampening apparent declines. This demonstrates an important need to understand post-breeding and overwinter survivorship and their influence on population regulation.

Qualitative Historical Assessment

CBC and BBS trend estimates suggest an 85–95% population decline over the past four decades, but it is important to determine if this decline represents only part of a longer historical process. Unfortunately, no quantitative surveys are available from earlier time periods. However, qualitative information on historical abundance illustrates how population status may have been changing even before CBC and BBS (Greenberg and Droege 1999). For example, several observers in the late 19th and early 20th centuries provided graphic descriptions of the high local abundance of this species during migration in the northern Great Plains, New England, and the Mississippi Valley (e.g., Beal 1890, Thompson 1891, J. H. Langille quoted in Beadslee and Mitchell 1965) that are not rivaled by any accounts since this period.

In addition, analysis of regional and state ornithological accounts indicates a long-term shift in descriptions of Rusty Blackbirds from "common

to abundant" to "rare or uncommon" (Fig. 9.2). Much of this change in the description of the species' status occurred prior to the recent 40-year decline documented by CBC and BBS. Analyses of a larger sample of local checklists show the same pattern in the qualitative descriptions of Rusty Blackbird abundance (Greenberg and Droege 1999). Such a long-term decline is more consistent with trends and patterns in winter habitat loss than with environmental change in the boreal forest, which is a more recent phenomenon. Interestingly, the pattern of change in the abundance descriptions for Rusty Blackbirds is similar between the Mississippi Alluvial Valley and the Southeastern Coastal Plain.

Past and Current Breeding Distribution

Another line of evidence for the decline of Rusty Blackbirds is the contraction of the breeding range, particularly in the southern boreal forest. Recent surveys from Canada suggest an apparent range contraction from southern areas traditionally occupied by Rusty Blackbirds. For example, recent surveys of 937 small wetlands in Alberta, Saskatchewan, and Manitoba have resulted in only 14 total detections (J. Morrisette, pers. comm.). Similar patterns have been observed in boreal uplands; in more than 20,000 point count surveys conducted across the western boreal forest in 1993–2006, only 80 locations documented Rusty

Figure 9.3. Range contraction of Rusty Blackbirds from the counties of eastern and northern Maine over the last century. Question marks indicate areas where knowledge of the current range remains anecdotal.

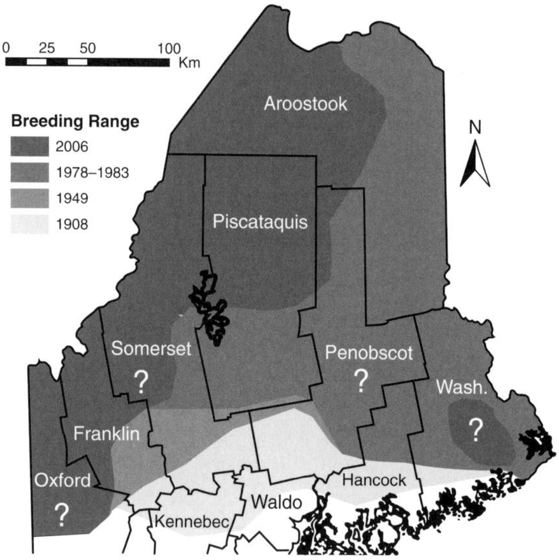

Blackbirds (S. Van Wilgenburg, J. Morrisette, and the Boreal Avian Modeling Project, http://www.borealbirds.ca). Additionally, several wetlands and lakes in eastern Saskatchewan and western Manitoba that were occupied in the 1970s are apparently no longer occupied (K. Hobson and A. Smith, unpubl. data).

The most complete analysis of range contraction comes from recent work in Maine, which suggests that the breeding range of Rusty Blackbird has contracted 65–160 km over the last century (Fig. 9.3). Early records (e.g., Knight 1908) report Rusty Blackbirds as summer residents throughout most of northern Maine and in southern parts of the state in Washington, Hancock, Waldo, and Kennebec counties. However, by 1950, Waldo, Kennebec, and southwestern Penobscot counties were not included in the Maine breeding range (Palmer 1949). The Maine Breeding Bird Atlas conducted from 1978–1983 confirmed the general distribution of breeding Rusty Blackbirds in Maine as described earlier by Palmer (Adamus 1987). Thus, although population declines may have occurred between the period of Palmer (1949) and the Maine atlas, a substantial range contraction during this time is not evident.

Two decades later, Hodgman and Hermann (2003) and L. Powell (unpubl. data) surveyed Rusty Blackbirds in Maine using call–response techniques at nearly 400 wetlands in core breeding areas in northern Aroostook, Piscataquis, and Somerset counties. Rusty Blackbirds were recorded at fewer than 10% of surveyed wetlands having apparently suitable habitat. Similar call–response surveys conducted from 2004 to 2005 at 350 wetlands in former breeding areas to the north and east did not yield a single confirmed detection (Hodgman and Yates 2007). Recent anecdotal evidence for Washington County suggests that perhaps only a few breeding pairs remain. These data and observations indicate that a substantial contraction in the breeding distribution of Rusty Blackbirds in Maine occurred between the 1980s and 2000.

Limited data from more northerly regions of the boreal forest suggest that range retractions or general declines may not have been as pronounced among these more remote areas. For example, resurveys of 45 of the original 61 wetland sites censused in 1975 as background data for the impact of a proposed gas pipeline in the Mackenzie Valley, Northwest Territories, suggested that wetland occupancy by Rusty Blackbirds had not changed substantially over the 30-year period (Machtans et al. 2007). This analysis suggests that even if local declines had occurred in the Northwest Territories, they are not of the magnitude that would be expected if the population had declined at a rate similar to that detected from Christmas Bird Count data (Machtans et al. 2007). Data from the Ontario Breeding Bird Atlas suggest substantial declines throughout the province except for the Hudson Bay Lowlands in the extreme north (Cadman et al. 2008).

HYPOTHESES FOR THE CAUSES
OF DECLINE

The reasons for the decline of the Rusty Blackbird are currently unknown. However, multiple factors operating at different spatial and temporal scales are likely responsible for the chronic and acute patterns of decline observed both regionally and range-wide. Greenberg and Droege (2003) propose three principal reasons, including: (1) winter habitat loss and degradation due to conversion and hydrologic alteration of bottomland hardwood habitats; (2) breeding habitat degradation due to logging, wetland drying, acidification, and mercury contamination; and (3) direct mortality associated with efforts to abate nuisance blackbird problems, particularly during winter. Each of these is discussed below along with possible predictions and tests that could help in evaluating their relative importance in explaining the observed declines.

Loss or Degradation of Winter Habitat

An estimated 75–80% of bottomland hardwood habitats in the United States have been converted to agriculture since European settlement (Hefner and Brown 1988, Hefner et al. 1994, Twedt and Loesch 1999). Certainly, the gross loss of wooded wetlands that has occurred in the southeastern United States would be consistent with a severe long-term decline in any species dependent on these habitats. The more difficult question is whether the pattern of decline indicated by BBS and CBC over the past 40 years is consistent with spatial patterns and rates of forested wetland loss during the same time period (Greenberg and Droege 1999).

For example, some low-lying and seasonally flooded regions of the Mississippi Alluvial Valley remained uncultivated until the 1970s because of low anticipated crop revenues and the high costs associated with converting these areas into production farmlands. By the 1970s, high soybean (*Glycine max*) prices enticed growers to convert and farm these areas, which may have been some of the best remaining wintering habitats for Rusty Blackbirds in the region. A distinct temporal pattern of land clearing from relatively drier contours more conducive to farming, to more flood-prone areas (Rudis 2001) likely had large implications for birds dependent on forested wetlands (Twedt et al.

2006). Subsequent declines in soybean prices, however, led to afforestation efforts, principally those undertaken as part of the U.S. Department of Agriculture Wetland Reserve and Conservation Reserve Programs (King et al. 2006). An estimated 162,000 ha were enrolled in the Wetland Reserve Program during 1990–2005 (Ducks Unlimited 2007). Considerable additional interest in afforestation has been spurred by utility industry investments in carbon sequestration programs that seek to offset carbon emissions while restoring cleared forests in an ecologically compatible manner (Caspersen et al. 2000, Houghton 2002, Shoch et al. 2009). This pattern of habitat loss followed by stabilized habitat trends and then a slow, steady gain appears to be consistent with the pattern seen in the CBC data, which show a precipitous drop in Rusty Blackbird abundance in the early 1970s. Emphasis of the Wetland Reserve and Conservation Reserve Programs on marginally productive croplands will result in afforestation activities that follow in reverse of earlier clearing patterns. This anticipated pattern suggests that as afforested lands mature into forests over the next several decades, considerable new habitat may become available for wintering Rusty Blackbird populations.

Nonetheless, this positive scenario of a net future return of Rusty Blackbird winter habitat needs to be kept in perspective. The great flexibility of the agricultural community to respond to changing markets, exemplified by earlier land clearing in response to rising soybean prices, suggests that markets for new commodities could quickly shift the habitat balance in the southeastern United States back again to agricultural lands. Currently, large shifts in crop production are occurring in response to new opportunities for biofuels such as corn (*Zea mays*) for ethanol. Overall biofuel production is increasing faster than 10% per year worldwide (Starke 2007), and corn production in the United States may double in the coming decades (Ringelman 2007). Some of this expansion is occurring at the expense of lands presently enrolled in wetlands conservation easements. Corn production has undergone recent expansion in the lower Mississippi Alluvial Valley in the past two years, but it is unclear how this will affect currently forested areas, or areas available for afforestation.

Another consideration is the management of existing wooded wetlands. As a primarily terrestrial, insectivorous species that forages in saturated or flooded soils, Rusty Blackbirds may be particularly sensitive to changes to natural forest flooding regimes resulting from drainage and diversions of water. In particular, the effects of the enormous water control projects of the Mississippi River and its major tributaries (Barry 1997) on Rusty Blackbirds is potentially quite large but has not been estimated. In much of the coastal Carolinas, wooded wetlands were historically impounded for rice production (Tompkins 1987). Many of these areas have regrown into second growth forest, but water levels are managed principally for waterfowl or are not managed at all. Management of water levels appropriate for blackbirds in both public and privately owned impoundments represents a significant opportunity for habitat enhancement and may also provide a means for research to examine how the depth, timing, duration, and spatial extent of inundation influence suitability of forested wetlands for Rusty Blackbirds. Thus far, only anecdotal natural history of the species suggests its sensitivity to the details of surface hydrology, and research is being undertaken to more rigorously test the impact of surface water conditions on the condition and survival of wintering blackbirds.

Loss or Degradation of Breeding Habitat

Until recently, the perception of the boreal forest as a vast expanse of undisturbed habitat has prevented many from invoking breeding habitat loss or degradation as an important factor in boreal forest bird declines. However, several widespread disturbances are noteworthy and deserve consideration as factors contributing to recent population reductions in Rusty Blackbirds and other species. Increasing surface air temperatures across Alaska and northwestern Canada have resulted in increases in the frequency, intensity, and extent of fire (Soja et al. 2006), as well as widespread drying of boreal wetlands (Klein et al. 2005, Riordan et al. 2006). In Alaska, the latter has resulted in a 19% loss of closed basin ponds, changes in water chemistry, decreases in macroinvertebrate abundance, and invasions of woody plants (Corcoran 2005, Klein et al. 2005, Riordan et al. 2006).

While the northern boreal forest feels the indirect effect of climate change, the southern boreal plains have long been impacted by direct human settlement and resource exploitation, often focused on wetland habitats. Large hydroelectric projects and concomitant reservoir development has led to the loss of riparian habitats and wetlands in several areas (Greenberg and Droege 1999). For example, over 1 million ha of forest was flooded in central Quebec (Gauthier and Aubry 1996). Furthermore, wetland changes have also been caused by the displacement of large volumes of underground and surface water during oil and gas extraction (Schmiegelow et al. 1997, Whitaker and Montevecchi 1999, Hobson et al. 2002, Bayne et al. 2005, Savignac 2006). Overall, the southern boreal forest has been impacted by timber harvest, agriculture, mining, and oil and gas development. Timber extraction is arguably the greatest threat to the overall integrity of the boreal forest. More than 60% of the commercially viable southern boreal forest has already been allocated to timber companies. However, other forms of land use have led to substantial habitat loss or degradation, particularly in the southern portion of the boreal plain. Eight percent of the boreal forest biome had been directly impacted by oil and gas extraction activities as of 2003 (Gauthier and Aubrey 1996). Approximately 79% of the forest plain ecozone at the southern edge of the boreal forest in Saskatchewan had been converted to agriculture since European settlement. Ongoing annual rates of deforestation in this ecozone across southern Canada range from 0.8 to 1.7% per year (Hobson et al. 2002). As a result of all this economic activity, the amount of intact habitat in the southern boreal forest is estimated to be no more than 75% of the pre-European coverage (Lee et al. 2006), and other estimates are much lower (Ricketts et al. 1999).

Boreal forests are dynamic habitats prone to natural disturbance regimes. More permanent habitat change can result from an interaction of natural forces and human management responses. For example, a massive outbreak of spruce budworms (*Choristoneura fumiferana*) from 1968 to 1985 defoliated balsam fir and spruce across 55 million ha of forest spanning the boreal zone from Lake Superior east to the Atlantic coast (Blais 1983, Bolgiano 2004). This and related salvage logging may have caused widespread changes to Rusty Blackbird breeding habitats.

The eastern boreal forest may be suffering a disproportionate impact from the fallout of industrial pollutants in heavily populated portions of the U.S. and Canada. Industrial pollution has decreased the quality of wetlands in the northeastern United States and eastern Canada by lowering pH, depleting environmental calcium (Greenberg and Droege 1999), and increasing concentrations of methylmercury (MeHg) (Lovett et al. 2009). Although aquatic ecosystems sensitive to environmental mercury loading are well established as having adverse impacts on piscivorous birds such as the Common Loon (*Gavia immer*) (Burgess and Meyer 2008; Evers et al. 2008), only recently has the availability of MeHg in wetland birds been identified as a major threat (Schwarzbach et al. 2006). Concentrations deemed to have adverse effects on egg hatchability, based on Heinz et al. (2008), have been documented for wild breeding populations of Icterids, including the Red-winged Blackbird (*Agelaius phoeniceus*) (Evers et al. 2005) and Rusty Blackbird (Evers, pers. comm.). Current sampling efforts of Rusty Blackbird tissues have documented significantly higher blood mercury concentrations in breeding versus wintering individuals, with highest levels recorded in the northeastern United States (BioDiversity Research Institute, unpubl. data). Reasons for elevated levels may be related to Rusty Blackbirds foraging on high trophic level invertebrates (e.g., arachnids; Cristol et al. 2008) from low-pH wetlands with frequent water level changes that are conducive to high methylation rates (Driscoll et al. 2007).

It should be noted that a number of other species that co-occur with Rusty Blackbirds in boreal wetlands during the breeding season are also suffering steep declines over the past few decades. These include Lesser Scaup (*Aythya affinis*); Black (*Melanitta nigra*), Scoters; Horned Grebe (*Podiceps auritus*), White-winged (*M. fusca*), and Surf (*M. perspicillata*); and Lesser Yellowlegs (*Tringa flavipes*) (Austin et al. 2000, Hannah 2004, North America Waterfowl Management Plan Committee 2004, Sauer et al. 2005, U.S. Fish and Wildlife Service 2006). Thus, Rusty Blackbirds may be responding to factors causing degradation in boreal wetlands that are having much broader impacts.

Blackbird Control Efforts

Rusty Blackbirds are not considered crop pests, but they do join other blackbirds and European Starlings (*Sturnus vulgaris*) in large communal winter roosts. Because of nuisance, property damage, and health concerns, winter roosts have been subjected to extensive control programs in the southeastern United States (Garner 1978, Heisterberg et al. 1987).

During 1974–1992, 83 roosts were sprayed with a surfactant (PA-14) which killed approximately 38 million blackbirds, principally Common Grackles (*Quiscalus quiscula*), European Starlings, Red-winged Blackbirds, and Brown-headed Cowbirds (*Molothrus ater*) (Dolbeer et al. 1997). Rusty Blackbirds have been estimated as comprising less than 1% of the birds in these mixed-species winter roosts (Meanley and Royall 1976). Thus, it was estimated that only 120,000 Rusty Blackbirds were affected by the PA-14 applications at the 83 winter roosts (Dolbeer et al. 1997). The use of the surfactant was discontinued when the EPA registration lapsed in 1992 and was not renewed. The period covered by this program coincides with that of the steepest decline as documented by the CBC data. The mortality estimates from control efforts do not suggest a magnitude that is equivalent to that of the global population declines (which would be on the order of millions of birds based on available trend and population estimates). Nonetheless, it would be instructive to examine the regional patterns of decline and how they relate to specific control events. Currently, the Rusty Blackbird–specific mortality estimates and the population estimates based on CBC appear to be insufficiently precise to support such an analysis.

The toxicant DCR-1339 (Starlicide®) is currently applied in rice-growing areas of Texas and Louisiana to reduce blackbird depredations to early-sprouting rice (Cummings and Avery 2003). Laced bait is applied on staging areas to affect birds entering and leaving large winter blackbird roosts. In parts of the Rusty Blackbird winter range, this toxicant is also used at feedlots and dairies for starling control (Homan et al. 2005). There is no estimate of the take of Rusty Blackbirds from DRC-1339 through either of these applications. Similarly, there is no information on how Rusty Blackbirds might be affected by legal removal of blackbirds as authorized in U.S. regulations under a blackbird depredation order (50 CFR 21.43) or by illegal shooting and trapping activities.

Migratory Allee Effect

Regardless of the external factors driving the decline of Rusty Blackbirds, two aspects of its life history may contribute to strong, negative density dependence at low population size (Allee effect). First, the species is often patchily distributed, following the local occurrence of appropriate wetland habitat across the landscape. Second, the migratory nature of the species means that individual birds settling to breed across the boreal landscape may not coalesce into viable breeding populations because of difficulty in locating mates or forming the loose nesting colonies observed in Alaska and Newfoundland (see below under Assessing Habitat Use on the Breeding Ground). A similar argument was made to account for the last stages of extinction of the Bachman's Warbler (*Vermivora bachmani*; Wilcove and Terborgh 1984). In particular, birds nesting in groups may be much more successful on a per capita basis than individual pairs due to group defense against nest predators, as has been found in Red-winged Blackbirds (Picman et al. 1988, Yasukawa et al. 1992). Taken together, the regional population size at which an extinction vortex (Gilpen and Soule 1986) is reached may be relatively high for migratory species, such as Rusty Blackbirds, compared to local resident species. This may explain rapid range contractions in regions, such as the southern boreal forest, where today seemingly appropriate breeding habitat is available but unoccupied.

LEGAL AND CONSERVATION STATUS OF THE RUSTY BLACKBIRD

Despite the evidence for a profound population decline, Rusty Blackbirds have only very recently received heightened conservation attention from governmental and private conservation entities. In the United States, Rusty Blackbirds receive the same legal protection afforded most migratory birds under the Migratory Bird Treaty Act (16 USC 703–711). The act establishes a federal prohibition against unauthorized pursuit, hunting, and killing of birds identified in various bilateral treaties between the United States and Great Britain (for Canada), Japan, and Russia. However, the act does not provide for the protection of habitats, nor does it mandate proactive conservation and management to sustain populations of protected species. Further, the act does allow for the permitted take of protected species when they threaten agricultural crops or human health or safety. In Canada, the Migratory Birds Convention Act (1994, c. 22, s. 19) implements the bilateral treaty between Canada and the United States for the protection of migratory birds. However, this act does not confer the same protections afforded blackbirds in the United States by the Migratory Bird Treaty

Act, largely because of the flexibility this affords the Canadian government in addressing crop depredation and other nuisance situations caused by blackbirds.

Until January 2011, the Rusty Blackbird could be legally taken in the United States without a permit under an existing depredation order for blackbirds, cowbirds, grackles, crows, and magpies (50 CFR 21.43). This order facilitates the lethal control of these birds when "committing or about to commit depredations . . . or when concentrated in such numbers and manner as to constitute a health hazard or other nuisance." In conjunction with all affected stakeholders, the U.S. Fish and Wildlife Service has excluded Rusty Blackbirds from this depredation order in acknowledgment of the increasing vulnerability of Rusty Blackbird populations to extirpation, the minimal threat they have on crops or other commodities, and their limited potential to congregate in numbers constituting a health or safety concern.

Non-regulatory designations add weight for the prioritization of funding and other resources dedicated to conservation, research, and monitoring of this species. Probably the most significant listing for the Rusty Blackbird is its recent inclusion on the IUCN Red List as a "vulnerable" species (IUCN 2007). NatureServe (2006) ranks Rusty Blackbirds as being secure at the range-wide scale, but notes vulnerabilities at state and provincial levels. Partners in Flight's North American Landbird Conservation Plan (Rich et al. 2004) lists Rusty Blackbird as a species of Continental Concern that is moderately abundant and widespread, but experiencing declines and high threats. The U.S. Fish and Wildlife Service has designated Rusty Blackbird as a Bird of Conservation Concern (U.S. Fish and Wildlife Service 2002), which earns the species additional attention and consideration in various activities of the agency. Similarly, the Committee on the Status of Endangered Wildlife in Canada recently identified Rusty Blackbird as a species of Special Concern (Savignac 2006).

Also outside of the regulatory arena was the formation of the International Rusty Blackbird Working Group in April 2005. This ad hoc group has consisted of approximately over 60 scientists, biologists, and program managers focused on (1) developing an overarching research and monitoring strategy to understand and reverse the species' decline; (2) providing information about the species and its decline to the greater scientific, conservation, and resource management community as well as to bird enthusiasts and the general public; and (3) serving as a forum for the real-time exchange of information among partners having a stake in Rusty Blackbird conservation. As of 2008, members of the working group had initiated research programs in both northern boreal habitats (particularly Alaska and Maine) and southeastern bottomland hardwood ecosystems (Mississippi, Arkansas, and South Carolina), supported by U.S. Fish and Wildlife Service, Department of Defense Legacy Program, U.S. Geological Survey, and the U.S. Forest Service. In 2008, efforts were initiated to collaborate with the eBird program (Cornell Laboratory of Ornithology and the National Audubon Society) to locate "hotspots" for the species during the winter and migration periods. All of these efforts are helping better describe the spatial and temporal distribution of Rusty Blackbirds as a basis for ongoing and planned research, monitoring, and conservation activities.

EFFORTS TO FILL CRITICAL INFORMATION GAPS

Habitat loss and degradation figure prominently in hypotheses for why the Rusty Blackbird is declining. Therefore, a complete understanding of habitat requirements and what features contribute to habitat quality is necessary to evaluate habitat-based hypotheses.

Habitat Use on the Breeding Ground

Understanding of the life history of the Rusty Blackbird has advanced surprisingly little since Bent (1958). In particular, information on habitat use on the boreal breeding grounds is scant and largely restricted to observations of limited numbers of birds in New England (Kennard 1920, Ellison 1990) and recent studies in Alaska (Corcoran 2006, Shaw 2006, P. Meyers, unpubl. data). Rusty Blackbirds are generally described as breeding solitarily and at low densities, but may also be found in small groups of a few to several pairs in Alaska and Newfoundland (Peters and Burleigh 1951, Gabrielson and Lincoln 1959, Ellison 1990), where the species appears to be most abundant (Sauer et al. 2005; Table 9.1). Rusty Blackbirds are patchily distributed and have often been reported occupying the same locations annually, with old

nests noted near active nests (Kennard 1920, Shaw 2006, R. Corcoran, pers. comm.). Recent studies in Alaska found nests as close as 75 m to one another but more typically >250 m apart. At sites with multiple pairs in Alaska, adults joined into groups of 3–7 individuals to mob potential predators near nests, particularly during the late nestling stage (Corcoran 2006, Shaw 2006, P. Meyers, pers. obs.).

The species breeds in bogs, wet meadows, or along ponds, lakes, and streams (Kennard 1920, Gabrielson and Lincoln 1959, Ellison 1990, Avery 1995, Sinclair et al. 2003). To many observers, the gestalt of a Rusty Blackbird breeding area includes a mixture of open water, flooded meadow or floating emergent vegetation, and conifers or tall shrubs. The dominant breeding habitat, however, varies among the principal regions where it has been studied. For example, Rusty Blackbirds are almost completely restricted to beaver ponds in New England, whereas in western Alaska they can be locally common in shrub and meadow vegetation along rivers, and in southern Alaska they are found along sloughs in early successional forests dominated by Sitka spruce (*Picea sitchensis*). Many observers have commented that Rusty Blackbirds can be seen flying great distances within or outside of the wetlands where their nests are located. Telemetry studies are needed to determine how extensive an area they require to forage during the breeding season and the nature of the overall habitat mosaic on which they depend.

Based on anecdotal accounts, it seems unlikely that the nest site itself is limiting the distribution of the species. This species builds a large nest low (<6 m) in small live or dead conifers (*Picea* or *Abies* spp.) or tall shrubs (*Salix* or *Alnus* spp.) (Kennard 1920, Gabrielson and Lincoln 1959, Ellison 1990, Sinclair et al. 2003, Shaw 2006). Although dependence on small conifers for nest sites is particularly marked in New England (Kennard 1920, Ellison 1990), the preferred tree or shrub species used for nesting appears to vary regionally. For example, along the drainage of the Yukon River and its tributaries in interior Alaska, over 75% of 37 nests were found in live or dead willows and often over water—fewer nests were found in live or dead spruce (*Picea glauca* and *P. mariana*; Corcoran 2006, Shaw 2006, K. Martin, unpubl. data, K. Sowl, unpubl. data). In contrast, in the coastal rainforest zone on the upper Copper River Delta, Alaska, 87% of 17 nests were found in Sitka spruce and the remainder in alder (P. Meyer, unpubl. data). It is likely that these regional patterns reflect differences in the availability of small trees rather than geographic variation in preference.

Winter Habitat Characteristics

As in other temperate zone blackbirds, Rusty Blackbirds have two distinct habitat needs during the non-breeding season: foraging areas and roosting sites. In terms of foraging areas, Rusty Blackbirds winter primarily in wet bottomland hardwood forests (Avery 1995). They are common in areas of continuous semi-flooded forest, but in drier areas they can be seen in association with smaller wooded wetlands, such as beaver ponds. They are also regularly found in more open habitats such as pecan (*Carya illinoiensis*) orchards and in forest fragments along creeks (Mettke-Hofmann unpubl. data). The extent to which Rusty Blackbirds use each of these habitats is poorly quantified, as is the effect that the use of these different habitats has on fitness. Within these habitats, birds show a preference for foraging near the edge of shallow water without regard for understory vegetation density or distance to forest edge. Habitat preference, in part, seemed to reflect the availability of preferred food. Rusty Blackbirds appeared to depend on two distinct dietary items: (1) small acorns and pecans, which are often eaten while associating with Common Grackles, whose large, strong bills are able to crack nutshells; and (2) invertebrates picked from water or soil, or captured after flipping leaf litter and floating vegetation. Consequently, Rusty Blackbirds appeared to select areas with the proper species of mast-producing oaks and hickories, or areas having a surface mosaic of water and moist soil that supported the appropriate invertebrate fauna. The qualitative information so far suggests that Rusty Blackbirds prefer forests with mature oaks (particularly willow oaks, *Quercus phellos*), areas with small creeks, and patchily inundated areas of shallow water and exposed substrate. Thus, overstory composition, the availability of mast-producing individuals of key oaks and hickories, the depth and extent of inundation, and general soil moisture regimes all appear to be important characteristics for future studies of winter habitat quality. The relative importance of these factors needs further evaluation.

In contrast to feeding areas, night roosts were more likely in fields than in scrub or forests. Preferred night roosts (20 to 400 birds) were often in afforestation areas with dense vegetation near the ground, but were also in fields with short vegetation, or trees or shrubs in residential yards—in the latter of which birds joined large roosts with other blackbird species. The preference for these relatively open, treeless habitats for night roosts is surprising given the avoidance of these habitats during feeding. Most of the trees in foraging areas are leafless in winter; birds may be quite conspicuous when using them as roosts. Thus, the ample low cover in new afforestation areas and vegetated fields and the dense foliage of evergreen trees in residential yards may give roosting blackbirds important protection against nocturnal predators and cold temperatures. Most Rusty Blackbirds are found either in single-species roosts or mixed with some Red-winged Blackbirds. However, as winter progresses, Rusty Blackbirds are more frequently found in large mixed-species blackbird roosts.

FUTURE RESEARCH

Now that it is understood that the Rusty Blackbird has suffered both a long-term and precipitous range-wide decline, future research needs to address three separate, but related needs: (1) a testing of predictions associated with the proposed causes of decline to begin to evaluate which factors are of paramount importance; (2) a plan consisting of concrete, proactive management recommendations to reverse the decline; and (3) continued monitoring of the population to provide both regional and global data to facilitate the first two objectives.

Testing Hypotheses for the Decline

Because resources are limited and the distribution of the species is vast, our ability to determine the causes of the decline will depend on the formation of specific hypothesis that can be tested by gathering information from well-defined, focal studies. The most important hypotheses are listed in Table 9.2, along with an attempt to define key data that will allow us to test key predictions from these hypotheses. Developing specific testable predictions for these various breeding-ground factors is difficult because of the lack of access to study this sparsely distributed species.

However, while a complete picture of the status of a widely distributed boreal population is impossible, evidence to evaluate the cause of decline could be developed with strategically placed studies. The site occupancy approach is being used for surveys on breeding areas in Maine (L. Powell, unpubl. data) and Alaska (S. Matsuoka and D. Shaw, unpubl. data) and, along with more detailed life history information, can help test specific breeding ground–related hypotheses. For example, because each of the listed hypotheses for decline has a distinct geographic pattern for the likelihood of impact, the local population trend and critical aspects of life history (particularly mating success, breeding success, nestling growth, or post-fledging survival) could be measured at key sites in the boreal region. If simple habitat loss or degradation was the driving force, then the populations should decline or contract away from areas where habitat has been destroyed or degraded. But if further environmental factors affect the quality of habitat, such as MeHg contamination or changes in food supply due to global warming and boreal wetland drying, then the reproductive output and particularly the growth, condition, and survival of young should be affected even in areas of seemingly appropriate habitat. For example, support for the effect of mercury contamination would be provided if (1) Rusty Blackbirds have declined disproportionately in areas that are most affected by mercury accumulation; (2) Rusty Blackbird blood and tissue show relatively high levels of mercury in birds from a region of greatest decline; and (3) reproductive anomalies associated with chronic mercury toxicity are documented. Similarly, acidification of wetlands should cause a distinct geographic pattern of decline associated with specific effects of calcium stress, such as eggshell thinning and poorly developed skeletons in young birds. Finally, if declines are disproportional in regions experiencing greater wetland drying or on a local level, smaller bodies of water might be more immediately affected, and the growth rate and fledging success of birds in these areas may be significantly lower than in less-affected areas (Table 9.2).

Similarly, the case for loss or degradation of winter habitat can be evaluated by exploring the correlation between population declines and land use changes at a finer scale than has been accomplished thus far. For a retrospective analysis,

exploration of individual CBC circles could shed light on the temporal and geographic patterns of decline in relation to available data on habitat change. In the future, focused surveying, using site occupancy approaches in different regions of the wintering ground, can be compared to habitat change as well. These abundance data, in conjunction with information on condition and other correlates of blackbird fitness, could be used to assess habitat quality, which can be added to predictive models for future decline. Other issues for ongoing and future research that will illuminate the nature of the Rusty Blackbird decline include:

- a more quantitative analysis of the loss of Rusty Blackbirds due to control efforts;

- sampling of Rusty Blackbirds for mercury and other contaminants in different portions of their breeding and winter range;

- comparisons of demographic variables, such as nest success and survivorship, for different regions of the breeding and wintering season;

- telemetry studies that reveal the details of how Rusty Blackbirds use the habitat mosaic for different aspects of their life history;

- further study of winter roosting behavior and how this affects access to preferred habitat and vulnerability to blackbird control;

- an examination of the possible role of disease and parasites in the decline of Rusty Blackbird populations; and

- more detailed studies of how forest management and hydrological interventions affect habitat quality, particularly on the wintering grounds.

Management Recommendations

An overall management strategy will have to await more definitive information on the causes of the Rusty Blackbird decline. However, many of the hypothesized causes for decline lend themselves to general conservation and environmental measures that are not exclusive to the Rusty Blackbird. For example, to the degree the development of boreal wetlands has contributed to the decline, efforts to protect and restore these ecosystems will contribute to the recovery of populations. Similarly, control of emissions that contribute to

TABLE 9.2

Possible causes, predictions, and necessary data to test predictions for declines in the Rusty Blackbird

Hypothesis	Prediction	Data Needed
	Breeding Season	
Breeding Habitat Loss	Steeper declines in areas/regions suffering greatest loss or conversion of habitat.	Temporal and spatial data on habitat loss, preferably that collected in association with monitoring data (e.g., habitat loss along survey routes).
Boreal Habitat Degradation		
Wetland drying	Steeper declines, lower fitness, and lower food availability in regions with the strongest apparent effect of drying (e.g., decreased surface water availability). Fitness correlates: low chick growth rate and condition (e.g., mass).	Detailed studies of site occupancy and reproductive performance across boreal wetland systems of differing size and varying vulnerability to drying.
Methyl mercury (MeHg) and wetland acidification (also includes wintering habitat conditions)	Steeper declines and lower fitness in regions with higher levels of MeHg contamination or acidification. High levels of MeHg in tissues from these regions. Fitness correlates: low adult survival and chick growth rates, high nest failure, and other reproductive anomalies. For acidification, egg shell thinning and skeletal deformities in chicks.	MeHg levels from areas with Rusty Blackbird surveys. Tissue samples from different boreal regions. Tissue samples from different winter regions if strong connection with breeding region can be established. Focused studies of reproductive success from wetlands with different levels of MeHg contamination.
Allee Effect	Local breeding populations extirpated as regional populations decline in the absence of any identifiable deterioration of habitat. As populations decline, more sites will be populated by unmated birds.	Detailed information on site occupancy, mating success, and reproductive success coupled with regional data on population trends.
	Non-breeding Season	
Non-breeding Habitat Loss	Abundance correlates with changes in area of de- and afforestation. Predictions can be refined as details of habitat quality are developed from studies.	Land use trend data to analyze with long-term data on blackbird abundance (e.g., land use within CBC circles). Bird responses to different habitat types in terms of condition and survival.

TABLE 9.2 (*continued*)

TABLE 9.2 (CONTINUED)

Hypothesis	Prediction	Data Needed
	Non-breeding Season	
Non-breeding Habitat Degradation: Changes in Forest Composition (mast-producing trees) and Hydrology	Within intact habitat, declines occurred in areas with greatest changes in the hydrology and composition of bottomland forests. Remaining concentrations are found in areas managed for ephemeral shallow water and high abundance of small acorn-producing trees.	Fine-scale data on the patterns of decline and the condition of bottomland hardwood forest. Atlas-level data on the distribution of winter concentrations of Rusty Blackbirds with appropriate habitat data.
Blackbird Control	Mortality from targeted and diffuse control efforts is sufficient to have an impact on population processes. Geographic and temporal pattern in declines correlates with major blackbird control efforts.	Number of Rusty Blackbirds taken by control efforts at large roosts. Systematic data on the geographic scope of blackbird control and the composition of roosts coupled with long-term, site-specific population trend data.

global warming, acid rain, and MeHg contamination may help restore Rusty Blackbirds to existing wetlands where they have disappeared. Similarly, policies and actions that protect and restore bottomland hardwood forests and other wooded wetlands would provide more habitat for wintering populations.

Certain more specific management recommendations may also be made proactively, based on the ongoing research focused on the habitat use and roosting behavior of the species. For example, Rusty Blackbirds appear to suffer indirectly from control efforts focused on blackbirds in general. Rusty Blackbirds often feed and roost in areas where relatively few individuals of other blackbirds occur. Control efforts that avoid prime habitat for Rusty Blackbird flocks and roosts would reduce the incidental mortality that might be occurring. Wooded wetlands are often managed for waterfowl and other wildlife. Managing the flooding regime of these areas to create appropriate foraging habitat may be another approach to increase the quality of protected habitats. Finally, afforestation efforts that include management for preferred food plants, such as oaks and hickories, for Rusty Blackbirds may also enhance habitat quality for recently reclaimed agricultural lands.

Ensuring that appropriate habitat is maintained in proximity to preferred nesting and roosting areas will require detailed understanding of how Rusty Blackbirds use the complex habitat mosaic associated with both boreal and southern wooded wetlands.

Monitoring

Because the Rusty Blackbird has a large and largely inaccessible breeding distribution throughout the boreal forest, population monitoring is likely to be based on the limited number of boreal BBS routes and the more geographically comprehensive, but less rigorously gathered, CBC data. It is unlikely in the near future that a more extensive monitoring program focused on Rusty Blackbirds will be implemented at either end of the annual cycle. It may be most strategic to incorporate more focused monitoring into studies aimed at testing the specific hypotheses for declines (see Testing Hypotheses for Declines). For example, regionally based citizen-science atlases or surveys using site occupancy approaches could be implemented in each of the BCRs that are most important in supporting wintering populations of Rusty Blackbirds. Such a survey effort could gather relatively unbiased information on the use of different habitats, which could later be correlated with land use change information. This effort could be made more efficient by concentrating on the late winter, when the populations are more stable and singing greatly increases the detectability of blackbird flocks. Increasing the breeding season coverage is even more problematic. However, since appropriate wetlands are often patchy and discrete, a program of conducting site occupancy surveys in a small number of representative regions seems within the realm of feasibility. Increasing the coverage of Rusty Blackbird surveys across the boreal forest region would be most feasible if incorporated with existing surveys for other types of birds, as is being done with waterfowl surveys in the Yukon Territory (P. Sinclair, unpubl. data) or could be done with surveys of boreal-nesting shorebirds (V. Johnston, unpubl. data).

ACKNOWLDGMENTS

The authors wish to thank our colleagues on the International Rusty Blackbird Working Group for continued discussions of the scientific and conservation issues involved in Rusty Blackbird conservation. The USDA Forest Service provided housing, cars, and logistical support. Furthermore, we thank Kathryn Heyden, Gerhard Hofmann, Catherine Ricketts, and Carl Smith for valuable field assistance during telemetry. The Yazoo National Wildlife Refuge (NWR) provided housing for our helpers. We thank the Theodore Roosevelt NWR Complex, the Northern Mississippi NWR Complex, Leroy Percy State Park, the Delta National Forest, the Delta Research and Extension Center, and Frederick Ballard and other individuals who granted permission to work on their refuges and lands. We are grateful to Robin Corcoran, Kate Martin, Paul Meyers, and Christine Sowl for providing unpublished data on nest sites and breeding behavior in Alaska. Jeff Wells supplied important information on the status of boreal forests. Financial support was provided by the U.S. Fish and Wildlife Service, the Canadian Wildlife Service, the Society for Tropical Ornithology, the Arthur-von-Gwinner Foundation, the Max Planck Institute for Ornithology, Friends of the National Zoo, and the Legacy Program of the Department of Defense.

LITERATURE CITED

Adamus, P. R. 1987. Atlas of breeding birds in Maine. Maine Department of Inland Fisheries and Wildlife, Augusta, ME.

Austin, J. E., A. D. Afton, M. G. Anderson, R. G. Clark, C. M. Custer, J. S. Lawrence, J. B. Pollard, and J. K. Ringelman. 2000. Declining scaup populations: issues, hypotheses, and research needs. Wildlife Society Bulletin 28:254–263.

Avery, M. L. 1995. Rusty Blackbird (*Euphagus carolinus*). A. Poole and F. Gill (editors), The birds of North America, No. 200. Academy of Natural Sciences, Philadelphia, and American Ornithologists' Union, Washington, D.C.

Barry, J. M. 1997. Rising tide: the great Mississippi flood of 1927 and how it changed America. Touchstone Books, New York, NY.

Bayne, E. M., S. Boutin, B. Tracz, and K. Charest. 2005. Functional and numerical responses of Ovenbirds (*Seiurus aurocapilla*) to changing seismic exploration practices in Alberta's boreal forest. Ecoscience 12:216–222.

Beadslee, C. S., and H. D. Mitchell. 1965. Birds of the Niagara frontier. Buffalo Society of Natural Sciences, Buffalo, NY.

Beal, F. E. 1890. Food of the bobolink, blackbirds, and grackles. USDA Biological Survey Bulletin 13.

Bent, A. C. 1958. Life histories of North American blackbirds, orioles, tanagers, and their allies. U.S. National National Museum Bulletin 211. Smithsonian Institution, Washington, DC.

Blais, J. R. 1983. Trends in the frequency, extent, and severity of spruce budworm outbreaks in eastern Canada. Canadian Journal of Forest Research 13:539–547.

Blancher, P. and J. V. Wells. 2005. The boreal forest region: North America's bird nursery. Canadian Boreal Initiative, Ottawa, ON, and Boreal Songbird Initiative, Seattle, WA.

Bolgiano, N. C. 2004. Cause and effect: changes in boreal bird irruptions in eastern North America relative to the 1970s spruce budworm infestation. American Birds 58:26–33.

Burgess, N. M., and M. W. Meyer. 2008. Methylmercury exposure associated with reduced productivity in Common Loons. Ecotoxicology 17:83–91.

Bystrak, D. 1981. The North American Breeding Bird Survey. Studies in Avian Biology 6:34–41.

Cadman, M. D., D. A. Sutherland, G. G. Beck, D. Lepage, and A. R. Couturier (editors). 2008. Atlas of the breeding birds of Ontario. Bird Studies Canada, Environment Canada, Ontario Field Ornithologists, Ontario Ministry of Natural Resources, and Ontario Nature.

Caspersen, J. P., S. W. Pacala, J. C. Jenkins, G. C. Hurtt, P. R. Moorcroft, and R. A. Birdsey. 2000. Contributions of land-use history to carbon accumulation in US forests. Science 290:1148–1151.

Corcoran, R. M. 2005. Lesser Scaup nesting ecology in relation to water chemistry and macroinvertebrates on the Yukon Flats, Alaska. M.S. thesis, University of Wyoming, Laramie, WY.

Corcoran, R. M. 2006. Nesting dynamics of Rusty Blackbirds on Innoko National Wildlife Refuge, Alaska: a preliminary study, spring 2006. U.S. Fish and Wildlife Service, Innoko National Wildlife Refuge, McGrath, AK.

Cristol, D. A., R. L. Brasso, A. M. Condon, R. E. Fovargue, S. L. Friedman, K. K. Hallinger, A. P. Monroe, and A. E. White. 2008. The movement of aquatic mercury through terrestrial food webs. Science 320:335.

Cummings, J. L., and M. L. Avery. 2003. An overview of current blackbird research in the southern rice growing region of the United States. Proceedings of the Wildlife Damage Management Conference 10:237–243.

Dolbeer, R. A., D. F. Mott, and J. L. Belant. 1997. Blackbirds and starlings killed at winter roosts from PA-14 applications, 1974–1992: implications for regional population management. Great Plains Wildlife Damage Control Workshop Proceedings 13:77–86.

Driscoll, C. T., Y. J. Han, C. Y. Chen, D. C. Evers, K. F. Lambert, T. M. Holsen, N. C. Kamman, and R. Munson. 2007. Mercury contamination in remote forest and aquatic ecosystems in the northeastern U.S.: Sources, transformations and management options. Bioscience 57:17–28.

Ducks Unlimited. 2007. Wetlands Reserve Program: helping to restore bottomland hardwoods in the Mississippi Alluvial Valley. Ducks Unlimited. <http://www.ducks.org/Conservation/Government Affairs/1622/WetlandsReserveProgram.html, July 2007>.

Ellison, W. G. 1990. The status and habitat of the Rusty Blackbird in Caledonia and Essex counties. Vermont Fish and Wildlife Department, Woodstock, VT.

Erskine, A. 1977. Birds in boreal Canada. Canadian Wildlife Service Report Series Number 41, Ottawa, ON.

Evers, D. C., L. Savoy, C. R. DeSorbo, D. Yates, W. Hanson, K. M. Taylor, L. Siegel, J. H. Cooley, M. Bank, A. Major, K. Munney, H. S. Vogel, N. Schoch, M. Pokras, M. W. Goodale, and J. Fair. 2008. Adverse effects from environmental mercury loads on breeding Common Loons. Ecotoxicology 17:69–81.

Evers, D. C., N. M. Burgess, L. Champoux, B. Hoskins, A. Major, W. M. Goodale, R. J. Taylor, R. Poppenga, and T. Daigle. 2005. Patterns and interpretation of mercury exposure in freshwater avian communities in northeastern North America. Ecotoxicology 14:193–221.

Gabrielson, I. N., and F. C. Lincoln. 1959. The birds of Alaska. Stackpole Co., Harrisburg, PA.

Garner, K. M. 1978. Management of blackbird and starling winter roost problems in Kentucky and Tennessee. Proceedings of the Vertebrate Pest Conference 8:54–59.

Gauthier, J., and Y. Aubry. 1996. Les Oiseaux nicheurs du: atlas nicheurs du Québec meridional. Association Québécoise des groupes d'ornithologues. Société Québécoise du protection des oiseaux. Canadian Wildlife Service, Environment Canada. Montreal, QC.

Gilpen, M. E., and M. E. Soule. 1986. Minimum viable populations: Processes of extinction. Pp. 19–34 in M. E. Soule (editor). Conservation biology, the science of scarcity and diversity. Sinauer Associates, Sunderland, MA.

Godfrey, W. E. 1986. The birds of Canada. Rev. ed. National Museums of Canada, Ottawa, ON.

Greenberg, R., and S. Droege. 1999. On the decline of the Rusty Blackbird and the use of ornithological literature to document long-term population trends. Conservation Biology 13:553–559.

Greenberg, R., and S. Droege. 2003. Rusty Blackbird: troubled bird of the boreal bog. Bird Conservation, June 2003.

Hannah, K. C. 2004. Status review and conservation plan for the Rusty Blackbird (*Euphagus carolinus*) in Alaska. Alaska Bird Observatory, Fairbanks, AK.

Hefner, J. M., and J. P. Brown. 1984. Wetland trends in southeastern U.S. Wetlands 4:1–11.

Hefner, J. M., B. O. Wilen, T. E. Dahl, and W. E. Frayer. 1994. Southeastern wetlands: status and trends, mid-1970s to mid-1980s. U.S. Fish and Wildlife Service and U.S. Environmental Protection Agency, Atlanta, GA.

Heinz, G. H., D. J. Hoffman, J. D. Klimstra, K. R. Stebbins, S. L. Kondrad, and C. A. Erwin. 2008. Species differences in the sensitivity of avian embryos to methylmercury. Archives of Environmental Contamination and Toxicology. doi:10.1007/s00244-008-9160-3.

Heisterberg, J. F., A. R. Stickley, Jr., K. M. Garner, and P. D. Foster, Jr. 1987. Controlling blackbirds and starlings at winter roosts using PA-14. Proceedings of the Eastern Wildlife Damage Control Conference 3:177–183.

Hobson, K. A., E. M. Bayne, and S. L. Van Wilgenburg. 2002. Large-scale conversion of forest to agriculture in the boreal plains of Saskatchewan. Conservation Biology 16:1530–1541.

Hodgman, T. P., and H. L. Hermann. 2003. Rusty Blackbird. Pp. 65–72 in H. L. Hermann, T. P. Hodgman, and P. deMaynadier (editors), A survey of rare, threatened, and endangered fauna in Maine: St. John uplands and boundary plateau (2001–2002). Maine Department of Inland Fisheries and Wildlife, Bangor, ME.

Hodgman, T. P., and D. Yates. 2007. Rusty Blackbird. Pp. 71–74 in G. M. Matula, Jr., T. P. Hodgman, and

P. de Maynadier (editors), A survey of rare, threatened, and endangered fauna in Maine: Aroostook Hills and lowlands (2003–2005). Maine Department of Inland Fisheries and Wildlife, Bangor, ME.

Homan, H. J., R. S. Stahl, J. J. Johnston, and G. M. Linz. 2005. Estimating DRC-1339 mortality using bioenergetics: a case study of the European Starling. Proceedings of the Wildlife Damage Management Conference 11:202–208.

Houghton, R. A. 2002. Magnitude, distribution and causes of terrestrial carbon sinks and some implications for policy. Climate Policy 2:71–88.

IUCN Red List. 2007. Birds on the IUCN red list. Bird Life International. <http://www.birdlife.org/action/science/species/global_species_programme/red_list.html> (July 2007).

Kennard, F. H. 1920. Notes on the breeding habits of the Rusty Blackbird in northern New England. Auk 37:412–422.

Kessel, B., and D. D. Gibson. 1978. Status and distribution of Alaska birds. Studies in Avian Biology 1:1–100.

King, S. L., D. J. Twedt, and R. R. Wilson. 2006. The role of the Wetland Reserve Program in conservation efforts in the Mississippi River Alluvial Valley. Wildlife Society Bulletin 34:914–920.

Klein, E., E. E. Berg, and R. Dial. 2005. Wetland drying and succession across the Kenai Peninsula lowlands, south-central Alaska. Canadian Journal of Forest Research 35:1931–1941.

Knight, O. W. 1908. Birds of Maine. Charles H. Glass & Co., Bangor, ME.

Lee, P., J. D. Gysbers, and Z. Stanojevic. 2006. Canada's forest landscape fragments: a first approximation (a Global Forest Watch Canada report). Global Forest Watch Canada, Edmonton, AB.

Link, W. A., and J. R. Sauer. 1996. Extremes in ecology: avoiding the misleading effects of sampling variation in summary analyses. Ecology 77:1633–1640.

Lovett, G. M., T. H. Tear, D. C. Evers, S. E. G. Findlay, B. J. Cosby, J. K. Dunscomb, C. T. Driscoll, and K. C. Weathers. 2009. Effects of air pollution on ecosystems and biological diversity in the eastern United States. Annals of the New York Academy of Sciences, 1162: 99–135.

Machtans, C. S., S. L. Van Wilgenburg, L. A. Armer, and K. A. Hobson. 2007. Retrospective comparison of the occurrence and abundance of Rusty Blackbird in the Mackenzie Valley, Northwest Territories. Avian Conservation and Ecology—Écologie et conservation des oiseaux 2(1):3. <http://www.ace-eco.org/vol2/iss1/art3/> (June 2007).

MacKenzie, D. I., J. D. Nichols, J. E. Hines, M. G. Knutson, and A. B. Franklin. 2003. Estimating site occupancy, colonization, and local extinction when a species is detected imperfectly. Ecology 84:2200–2207.

MacKenzie, D. I., J. D. Nichols, G. B. Lachman, S. Droege, J. A. Royle, and C. A. Langtimm. 2002. Estimating site occupancy rates when detection probabilities are less than one. Ecology 83:2248–2255.

Meanley, B., and W. C. Royall, Jr. 1976. Nationwide estimates of blackbirds and starlings. Proceedings of the Bird Control Seminar 7:39–40.

NatureServe. 2006. NatureServe explorer: an online encyclopedia of life, version 6.1. NatureServe, Arlington, VA. <http://www.natureserve.org/explorer> (June 2007).

Niven, D. K., J. R. Sauer, G. S. Butcher, and W. A., Link. 2004. Christmas bird count provides insights into population change in land birds that breed in the boreal forest. American Birds 58:10–20.

North American Waterfowl Management Plan Committee. 2004. 2004 North American waterfowl management plan 2004. Strategic guidance: strengthening the biological foundation. Canadian Wildlife Service, U.S. Fish and Wildlife Service, Secretaria de Medio Ambiente y Recursos Naturales.

Palmer, R. S. 1949. Maine birds. Bulletin of the Museum of Comparative Zoology, Harvard College, Vol. 102. Cambridge, MA.

Peters, H. S., and T. D. Burleigh. 1951. The birds of Newfoundland. Houghton Mifflin Co., Boston, MA.

Picman, J., M. Leonard, and A. Horn. 1988. Antipredation role of clumped nesting by marsh-nesting Red-winged Blackbirds. Behavioral Ecology and Sociobiology 22:9–15.

Rappole, J. H., and A. R. Tipton. 1991. New harness design for attachment of radio transmitters to small passerines. Journal of Field Ornithology 62:335–337.

Rich, T. D., C. J. Beardmore, H. Berlanga, P. J. Blancher, M. S. W. Bradstreet, G. S. Butcher, D. W. Demarest, E. H. Dunn, W. C. Hunter, E. E. Inigo-Elias, J. A. Kennedy, A. M. Martell, A. O. Panjabi, D. N. Pashley, K. V. Rosenberg, C. M. Rustay, J. S. Wendt, and T. C. Will. 2004. Partners in Flight North American Landbird Conservation Plan. Cornell Lab of Ornithology, Ithaca, NY.

Ricketts, T. H., E. Dinerstein, D. M. Olson, C. J. Loucks, W. Eichbaum, D. DellaSala, K. Kavanagh, P. Hedao, P. T. Hurley, K. M. Carney, R. Abell, and S. Walters. 1999. Terrestrial ecoregions of North America: a conservation assessment. Island Press, Washington, DC.

Ringelman, J. 2007. Biofuels and ducks. <http://www.ducks.org/DU_Magazine/DUMagazineMayJune2007/3213/BiofuelsandDucks.html>.

Riordan, B., D. Verbyla, and A. D. McGuire. 2006. Shrinking ponds in subarctic Alaska based on

1950–2002 remotely sensed images. Journal of Geophysical Research 111, G04002.

Rosenberg, K. V., and P. J. Blancher. 2005. Setting numerical population objectives for priority landbird species. Pp. 57–67 *in* C. J. Ralph and T. D Rich (editors), Bird conservation and implementation in the Americas: proceedings of the third international Partners in Flight Conference. USDA Forest Service General Technical Report PSW-GTR-191. USDA Forest Service, Pacific Southwest Research Station, Albany, CA.

Rudis, V. A. 2001. Composition, potential old growth, fragmentation, and ownership of Mississippi Alluvial Valley bottomland hardwoods: a regional assessment of historic change. Pp. 28–48 *in* P. B. Hamel and T. L. Foti (editors), Bottomland hardwoods of the Mississippi Alluvial Valley: characteristics and management of natural function, structure, and composition. General Technical Report SRS-42. U.S. Forest Service, Asheville, NC.

Sauer, J. R., J. E. Hines, and J. Fallon. 2005. The North American Breeding Bird Survey, results and analysis 1966–2005, version 6.2.2006. USGS Patuxent Wildlife Research Center, Laurel, MD. <http://www.mbr-pwrc.usgs.gov/bbs/bbs.html> (June 2007).

Savignac, C. 2006. COSEWIC assessment and status report on the Rusty Blackbird (*Euphagus carolinus*) in Canada. Committee on the Status of Endangered Wildlife in Canada, Ottawa, ON.

Schmiegelow, F. K. A., C. S. Machtans, and S. J. Hannon. 1997. Are boreal birds resilient to forest fragmentation? an experimental study of short-term community responses. Ecology 78:1914–1932.

Schwarzbach, S. E., J. D. Albertson, and C. M. Thomas. 2006. Effects of predation, flooding, and contamination on reproductive success of California Clapper Rails (*Rallus longirostris obsoletus*) in San Francisco Bay. Auk 123:45–60.

Shaw, D. 2006. Breeding ecology and habitat affinities of an imperiled species, the Rusty Blackbird (*Euphagus carolinus*), in Fairbanks, Alaska: preliminary results. Alaska Bird Observatory, Fairbanks, AK.

Shoch, D. T., G. Kaster, A. Hohl, and R. Souter. 2009. Carbon storage of bottomland hardwood afforestation in the Lower Mississippi Valley, USA. Wetlands 29:535–542

Sinclair, P. H., W. A. Nixon, C. D. Eckert, and N. L. Hughes. 2003. Birds of the Yukon. University of British Columbia Press, Vancouver, BC.

Soja, A. J., N. M. Tchebakova, N. H. F. French, M. D. Flannigan, H. H. Shugart, B. J. Stocks, A. I. Sukhinin, E. I. Parfenova, F. S. Chapin III, and P. W. Stackhouse, Jr. 2006. Climate-induced boreal forest change: predications versus current observations. Global and Planetary Change 54:274–296.

Starke, L. (editor). 2007. State of the world: our urban future. W. W. Norton Co., New York, NY.

Stocks, B. J., M. A. Fosberg, T. J. Lynham, L. Mearns, B. M. Wotton, Q. Yang, J.-Z. Jin, K. Lawrence, G. R. Hartley, J. A. Mason, and D. W. McKenney. 2004. Climate change and forest fire potential in Russian and Canadian boreal forests. Climate Change 38:1–13.

Thogmartin, W. E., F. P. Howe, F. C. James, D. H. Johnson, E. T. Reed, J. R. Sauer, and F. R. Thompson III. 2006. A review of the population estimation approach of the North American Landbird Conservation Plan. Auk 123:892–904.

Thompson, E. E. 1891. The birds of Manitoba. Proceedings of the U.S. National Museum, Vol. 13, no. 841. Washington, DC.

Thompson, W. L. (editor). 2004. Sampling rare and elusive species: concepts, designs, and techniques for estimating population parameters. Island Press, Washington, DC.

Tompkins, M. E. 1987. Scope and status of coastal wetlands impoundments. Pages 31–57 *in* M. R. Devoe and B. S. Baughman (editors), Impoundments: ecological characterization, management, status, and use, Vol. 2. South Carolina Sea Grant Consortium, Charleston, SC.

Twedt, D. J., and C. R. Loesch. 1999. Forest area and distribution in the Mississippi Alluvial Valley: implications for breeding bird conservation. Journal of Biogeography 26:1215–1224.

Twedt, D. J., W. B. Uihlein III, and A. B. Elliott. 2006. A spatially explicit decision support model for restoration of forest bird habitat use. Conservation Biology 20:100–110.

U. S. Fish and Wildlife Service. 2002. Birds of conservation concern 2002. Division of Migratory Bird Management, Arlington, VA. <http://www.fws.gov/migratorybirds/reports/BCC2002.pdf> (June 2007).

U. S. Fish and Wildlife Service. 2006. Waterfowl population status, 2006. U. S. Department of the Interior, Washington, DC.

Whitaker, D. M., and W. A. Montevecchi. 1997. Breeding bird assemblage associated with riparian, interior forest, and non-riparian edge habitats in a balsam fir ecosystem. Canadian Journal of Zoology 27:1159–1167.

Whitaker, D. M., and W. A. Montevecchi. 1999. Breeding bird assemblages inhabiting riparian buffer strips in Newfoundland, Canada. Journal of Wildlife Management 63:167–179.

Wilcove, D. S., and J. W. Terborgh. 1984. Patterns of population decline in birds. American Birds 38:10–13.

Yasukawa, K., R. A. Boley, J. L. McClure, and J. Zanocco 1992. Nest dispersion in the Red-winged Blackbird. Condor 94:775–777.

Gull, Mew (*Larus canus*), 11, 18
Gull, Ring-billed (*Larus delawarensis*), 18, 103
Gull, Ross's (*Rhodostethia rosea*), 19
Gull, Sabine's (*Xema sabini*), 19, 103
Gull, Thayer's (*Larus thayeri*), 18
Gyrfalcon (*Falco rusticolus*), 18

habitat fragmentation, 30–32
Harrier, Northern (*Circus cyaneus*), 18, 102
harvest mortality, 28
    sustainable rate in sea ducks, 58
Hawk, Broad-winged (*Buteo platypterus*), 12, 18, 102
    wintering in Tropical Andes, 97, 99
Hawk, Cooper's (*Accipiter cooperii*), 18
Hawk, Red-shouldered (*Buteo lineatus*), 18
Hawk, Red-tailed (*Buteo jamaicensis*), 18, 102
Hawk, Rough-legged (*Buteo lagopus*), 13, 18
Hawk, Sharp-shinned (*Accipiter striatus*), 12, 18
Hawk, Swainson's (*Buteo swainsoni*), 18, 102
Heron, Great Blue (*Ardea herodias*), 18, 102
Heron, Green (*Butorides virescens*), 102
herring, Pacific (*Clupea pallasi*), 56
Hummingbird, Ruby-throated (*Archilochus colubris*), 19
Hummingbird, Rufous (*Selasphorus rufus*), 19
hydropower
    habitat loss, 2, 114

Important Bird Areas, 8, 69, 71, 95–105
    conservation role for boreal migrants, 100
    description, 96
    political support in Tropical Andes, 100
    role of abundance data in identification, 69
    wintering sites for boreal migrants, 95–105
indigenous communities land-use plans, 3
International Rusty Blackbird Technical Working Group, 116–117
isotope ratios, 55

Jaeger, Long-tailed (*Stercorarius longicaudus*), 18, 103
Jaeger, Parasitic (*Stercorarius parasiticus*), 18, 103
Jaeger, Pomarine (*Stercorarius pomarinus*), 18
Jay, Blue (*Cyanocitta cristata*), 19
Jay, Gray (*Perisoreus canadensis*), 10–11, 13, 19, 80
Junco, Dark-eyed (*Junco hyemalis*), 11, 13, 20, 74, 81
    relative abundance in northern Ontario, Canada, 68 (fig.)
Junco, "Cassiar's" Dark-eyed (*Junco hyemalis cismontanus*), 75

Kestrel, American (*Falco sparverius*), 18
Killdeer (*Charadrius vociferus*), 18, 102
Kingbird, Eastern (*Tyrannus tyrannus*), 19, 104
Kingfisher, Belted (*Megaceryle alcyon*), 12, 19, 103
Kinglet, Golden-crowned (*Regulus satrapa*), 12, 20
Kinglet, Ruby-crowned (*Regulus calendula*), 11, 20, 74, 80, 91
Knot, Red (*Calidris canutus*), 13, 18, 103
kriging, 66, 70

Lark, Horned (*Eremophila alpestris*), 19
Long Point Bird Observatory, 75–77

Longspur, Lapland (*Calcarius lapponicus*), 13, 20
Longspur, Smith's (*Calcarius pictus*), 12–13, 20
Loon, Common (*Gavia immer*), 11, 17, 115
Loon, Pacific (*Gavia pacifica*), 11, 17
Loon, Red-throated (*Gavia stellata*), 17
Loon, Yellow-billed (*Gavia adamsii*), 17

Magpie, Black-billed (*Pica hudsonia*), 12, 19
Maine Breeding Bird Atlas, 112
Mallard (*Anas platyrhynchos*), 17, 26, 39–40, 42
management actions, 1
    adaptive management approach, 32–34
    climate change, 31
    ecological benchmarks, 32
    flooding regimes and Rusty Blackbirds, 113–114
Martin, Purple (*Progne subis*), 19, 104
Meadowlark, Eastern (*Sturnella magna*), 21
Meadowlark, Western (*Sturnella neglecta*), 21
Merganser, Common (*Mergus merganser*), 11, 17, 26, 39–40
Merganser, Hooded (*Lophodytes cucullatus*), 11, 17, 26, 39–40
Merganser, Red-breasted (*Mergus serrator*), 12, 17, 26, 39–40
Merlin (*Falco columbarius*), 11, 18, 102
methylmercury
    threat to wetland bird species, 114–115
migrants. *See also* migration, migratory funneling
    boreal component of migrant bird communities in eastern North America, 73–83
    conservation status in Tropical Andes, 102–105 (table)
    diet shift, 91
    ecological role in wintering regions, 90–91
    estimation of diversity and abundance in Neotropics, 86–87
    neotropical winter bird communities, 85–93
    non-boreal breeding migrants, 10
    population trends, 96
    role as pollinators and seed dispersers on wintering areas, 91
    species richness and abundance in Neotropics by country, 87–88
    species richness and Bird Conservation Regions, 88–89
    species richness, abundance and density of wintering birds, 88–89
    winter bird communities, 85–93
    wintering in Tropical Andes, 95–105
migration monitoring network, 77–78
migration. *See also* migrants, migratory funneling
    band recoveries of boreal species, 74
    ice break-up, 58
    importance of boreal forest, 10
    movement pattern of boreal species, 74, 77
    percent boreal birds at fall banding locations in eastern North America, 76
    weather, 58
Migratory Bird Treaty Act, 116
Migratory Birds Convention Act, 116
migratory connectivity, 42, 56, 87

Redhead (*Aythya americana*), 17, 39–40
Redpoll, Common (*Acanthis flammea*), 12, 21
Redpoll, Hoary (*Acanthis hornemanni*), 21
Redstart, American (*Setophaga ruticilla*), 12, 20, 104
remote sensing, 45–46
roads, 1–2, 33
    ecological effects, 30–32
Robin, American (*Turdus migratorius*), 12, 20
Rosy-Finch, Gray-crowned (*Leucosticte tephrocotis*),
    11, 21

Salix spp., 117
Sanderling (*Calidris alba*), 13, 18, 103
Sandpiper, Baird's (*Calidris bairdii*), 13, 18, 103
Sandpiper, Buff-breasted (*Tryngites subruficollis*),
    13, 18, 103
Sandpiper, Least (*Calidris minutilla*), 11, 18, 103
    wintering in Tropical Andes, 97
Sandpiper, Pectoral (*Calidris melanotos*), 10, 13, 18, 103
Sandpiper, Purple (*Calidris maritima*), 18
Sandpiper, Rock (*Calidris ptilocnemis*), 18
Sandpiper, Semipalmated (*Calidris pusilla*), 12–13, 18,
    100, 103
Sandpiper, Solitary (*Tringa solitaria*), 11, 13, 18, 102
    wintering in Tropical Andes, 97
Sandpiper, Spotted (*Actitis macularius*), 11, 18, 102
    wintering in Tropical Andes, 97
Sandpiper, Stilt (*Calidris himantopus*), 12–13, 18, 103
Sandpiper, Upland (*Bartramia longicauda*), 18, 102
    wintering in Tropical Andes, 97
Sandpiper, Western (*Calidris mauri*), 18, 103
Sandpiper, White-rumped (*Calidris fuscicollis*),
    10, 13, 18, 103
Sapsucker, Red-breasted (*Sphyrapicus ruber*), 19
Sapsucker, Red-naped (*Sphyrapicus nuchalis*), 19
Sapsucker, Yellow-bellied (*Sphyrapicus varius*),
    11, 19, 80, 103
satellite telemetry, 44
Scaup, Greater (*Aythya marila*), 10–11, 17, 26, 39–40
Scaup, Lesser (*Aythya affinis*), 11, 17, 26, 31, 39–40,
    58, 102
    nest failure, 56
    population decline, 115
science priorities
    Breeding Bird Atlas projects and relative abundance
        data, 69
    carrying capacity, 29
    climate change and human disturbance, 31
    conceptual framework for research, 32, 65
    distribution and abundance information in northern
        boreal extent, 70
    disturbance threats to waterfowl and wetlands, 26
    ecological role of boreal migrants in Neotropics, 86, 91
    geographical source areas of migrants, 77
    habitat preferences of boreal migrants in Tropical
        Andes, 99–100
    impact of boreal forest conversion to agriculture, 30
    monitoring populations of boreal forest-nesting
        birds, 74
    protected areas and neotropical migrants, 96
    road effects and long-term monitoring, 31

Rusty Blackbird research needs, 118–122
waterfowl densities and landscape features, 28
waterfowl information needs, 34
wetland mapping, 28
Scoter, Black (*Melanitta americana*), 11, 13, 17, 26, 39–40
    population decline, 115
Scoter, Surf (*Melanitta perspicillata*), 11, 13, 17, 26, 39–63
    anthropogenic threats to wintering areas, 42
    breeding biology, 42, 44, 53, 55–56
    breeding range, 43 (fig.), 49, 53, 55
    conservation threats, 58
    migration distance, 43
    movement among wintering areas, 56
    nesting area fidelity, 49–50, 55–56
    nesting area selection, 56–58
    nesting site characteristics, 51–52
    population estimate and decline, 42, 115
    wintering origin and settlement on breeding areas,
        51, 55–56
Scoter, White-winged (*Melanitta fusca*), 10, 11, 13, 17, 26,
    39–40, 55, 57
    population decline, 115
Shoveler, Northern (*Anas clypeata*), 12, 17, 39–40, 102
Shrike, Loggerhead (*Lanius ludovicianus*), 19
Shrike, Northern (*Lanius excubitor*), 10–11, 13, 19, 80
Siskin, Pine (*Spinus pinus*), 12, 21
Snipe, Wilson's (*Gallinago delicata*), 11, 18
Solitaire, Townsend's (*Myadestes townsendi*), 20
Sora (*Porzana carolina*), 11, 18, 102
southern boreal forest, 66–67, 70, 107, 111
    annual rates of deforestation, 114
    anthropogenic factors, 114–115
    conservation status, 1
    population monitoring bias, 74
soybean (*Glycine max*), 113
Sparrow, American Tree (*Spizella arborea*), 12, 20
Sparrow, Baird's (*Ammodramus bairdii*), 20
Sparrow, Brewer's (*Spizella breweri*), 20
Sparrow, Chipping (*Spizella passerina*), 12, 20
Sparrow, Clay-colored (*Spizella pallida*), 11, 20, 81
Sparrow, Fox (*Passerella iliaca*), 11, 20, 81
Sparrow, Golden-crowned (*Zonotrichia atricapilla*),
    11, 20
Sparrow, Harris's (*Zonotrichia querula*), 13, 20
Sparrow, House (*Passer domesticus*), 21
Sparrow, Le Conte's (*Ammodramus leconteii*), 11, 20, 81
Sparrow, Lincoln's (*Melospiza lincolnii*), 11, 13, 20, 77, 81
Sparrow, Nelson's Sharp-tailed (*Ammodramus nelsoni*),
    12, 20
Sparrow, Savannah (*Passerculus sandwichensis*), 12, 20
Sparrow, Song (*Melospiza melodia*), 20
Sparrow, Swamp (*Melospiza georgiana*), 11, 20, 81
Sparrow, Vesper (*Pooecetes gramineus*), 20
Sparrow, White-crowned (*Zonotrichia leucophrys*),
    11, 20, 81
Sparrow, "Eastern" White-crowned, (*Zonotrichia
    leucophrys leucophrys*), 81
Sparrow, "Gambel's" White-crowned (*Zonotrichia
    leucophrys gambelii*), 81
Sparrow, White-throated (*Zonotrichia albicollis*), 11, 13,
    20, 74, 81

1. Kessel, B., and D. D. Gibson. 1978.
Status and Distribution of Alaska Birds.

2. Pitelka, F. A., editor. 1979.
Shorebirds in Marine Environments.

3. Szaro, R. C., and R. P. Balda. 1979.
Bird Community Dynamics in a Ponderosa Pine Forest.

4. DeSante, D. F., and D. G. Ainley. 1980.
The Avifauna of the South Farallon Islands, California.

5. Mugaas, J. N., and J. R. King. 1981.
Annual Variation of Daily Energy Expenditure by the Black-billed Magpie: A Study of Thermal and Behavioral Energetics.

6. Ralph, C. J., and J. M. Scott, editors. 1981.
Estimating Numbers of Terrestrial Birds.

7. Price, F. E., and C. E. Bock. 1983.
Population Ecology of the Dipper (Cinclus mexicanus) *in the Front Range of Colorado.*

8. Schreiber, R. W., editor. 1984.
Tropical Seabird Biology.

9. Scott, J. M., S. Mountainspring, F. L. Ramsey, and C. B. Kepler. 1986.
Forest Bird Communities of the Hawaiian Islands: Their Dynamics, Ecology, and Conservation.

10. Hand, J. L., W. E. Southern, and K. Vermeer, editors. 1987.
Ecology and Behavior of Gulls.

11. Briggs, K. T., W. B. Tyler, D. B. Lewis, and D. R. Carlson. 1987.
Bird Communities at Sea off California: 1975 to 1983.

12. Jehl, J. R., Jr. 1988.
Biology of the Eared Grebe and Wilson's Phalarope in the Nonbreeding Season: A Study of Adaptations to Saline Lakes.

13. Morrison, M. L., C. J. Ralph, J. Verner, and J. R. Jehl, Jr., editors. 1990.
Avian Foraging: Theory, Methodology, and Applications.

14. Sealy, S. G., editor. 1990.
Auks at Sea.

15. Jehl, J. R., Jr., and N. K. Johnson, editors. 1994.
A Century of Avifaunal Change in Western North America.

16. Block, W. M., M. L. Morrison, and M. H. Reiser, editors. 1994.
The Northern Goshawk: Ecology and Management.

17. Forsman, E. D., S. DeStefano, M. G. Raphael, and R. J. Gutiérrez, editors. 1996.
Demography of the Northern Spotted Owl.

18. Morrison, M. L., L. S. Hall, S. K. Robinson, S. I. Rothstein, D. C. Hahn, and T. D. Rich, editors. 1999.
Research and Management of the Brown-headed Cowbird in Western Landscapes.

19. Vickery, P. D., and J. R. Herkert, editors. 1999.
Ecology and Conservation of Grassland Birds of the Western Hemisphere.

20. Moore, F. R., editor. 2000.
Stopover Ecology of Nearctic–Neotropical Landbird Migrants: Habitat Relations and Conservation Implications.

21. Dunning, J. B., Jr., and J. C. Kilgo, editors. 2000.
Avian Research at the Savannah River Site: A Model for Integrating Basic Research and Long-term Management.

22. Scott, J. M., S. Conant, and C. van Riper, II, editors. 2001.
Evolution, Ecology, Conservation, and Management of Hawaiian Birds: A Vanishing Avifauna.

23. Rising, J. D. 2001.
Geographic Variation in Size and Shape of Savannah Sparrows (Passerculus sandwichensis).

24. Morton, M. L. 2002.
The Mountain White-crowned Sparrow: Migration and Reproduction at High Altitude.

25. George, T. L., and D. S. Dobkin, editors. 2002. *Effects of Habitat Fragmentation on Birds in Western Landscapes: Contrasts with Paradigms from the Eastern United States.*

26. Sogge, M. K., B. E. Kus, S. J. Sferra, and M. J. Whitfield, editors. 2003. *Ecology and Conservation of the Willow Flycatcher.*

27. Shuford, W. D., and K. C. Molina, editors. 2004. *Ecology and Conservation of Birds of the Salton Sink: An Endangered Ecosystem.*

28. Carmen, W. J. 2004. *Noncooperative Breeding in the California Scrub-Jay.*

29. Ralph, C. J., and E. H. Dunn, editors. 2004. *Monitoring Bird Populations Using Mist Nets.*

30. Saab, V. A., and H. D. W. Powell, editors. 2005. *Fire and Avian Ecology in North America.*

31. Morrison, M. L., editor. 2006. *The Northern Goshawk: A Technical Assessment of Its Status, Ecology, and Management.*

32. Greenberg, R., J. E. Maldonado, S. Droege, and M. V. McDonald, editors. 2006. *Terrestrial Vertebrates of Tidal Marshes: Evolution, Ecology, and Conservation.*

33. Mason, J. W., G. J. McChesney, W. R. McIver, H. R. Carter, J. Y. Takekawa, R. T. Golightly, J. T. Ackerman, D. L. Orthmeyer, W. M. Perry, J. L. Yee, M. O. Pierson, and M. D. McCrary. 2007. *At-Sea Distribution and Abundance of Seabirds off Southern California: A 20-Year Comparison.*

34. Jones, S. L., and G. R. Geupel, editors. 2007. *Beyond Mayfield: Measurements of Nest-Survival Data.*

35. Spear, L. B., D. G. Ainley, and W. A. Walker. 2007. *Foraging Dynamics of Seabirds in the Eastern Tropical Pacific Ocean.*

36. Niles, L. J., H. P. Sitters, A. D. Dey, P. W. Atkinson, A. J. Baker, K. A. Bennett, R. Carmona, K. E. Clark, N. A. Clark, C. Espoz, P. M. González, B. A. Harrington, D. E. Hernández, K. S. Kalasz, R. G. Lathrop, R. N. Matus, C. D. T. Minton, R. I. G. Morrison, M. K. Peck, W. Pitts, R. A. Robinson, and I. L. Serrano. 2008. *Status of the Red Knot* (Calidris canutus rufa) *in the Western Hemisphere.*

37. Ruth, J. M., T. Brush, and D. J. Krueper, editors. 2008. *Birds of the US–Mexico Borderland: Distribution, Ecology, and Conservation.*

38. Knick, S. T., and J. W. Connelly, editors. 2011. *Greater Sage-Grouse: Ecology and Conservation of a Landscape Species and Its Habitats.*

39. Sandercock, B. K., K. Martin, and G. Segelbacher, editors. 2011. *Ecology, Conservation, and Management of Grouse.*

40. Forsman, E. D., et al. 2011. *Population Demography of Northern Spotted Owls.*

41. Wells, J. V., editor. 2011. *Boreal Birds of North America: a hemispheric view of their conservation links and significance.*

Indexer:	Leslie A. Robb
Composition:	MPS Limited, a Macmillan Company
Text:	Scala
Display:	Scala Sans
Printer and Binder:	Thomson-Shore

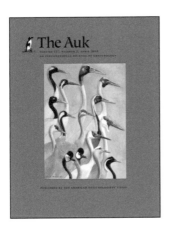